# Gravitation, Electromagnetism and Cosmology: toward a new synthesis

Edited by Konrad Rudnicki

Apeiron
Montreal

Published by C. Roy Keys Inc.
4405, rue St-Dominique
Montreal, Quebec H2W 2B2 Canada
*http://redshift.vif.com*



First Published 2001

*National Library of Canada Cataloguing in Publication Data*

Main entry under title:

Gravitation, electromagnetism and cosmology: toward a new synthesis

Based on papers originally presented at an international conference on Redshifts and gravitation in a relativistic universe, held Sept. 17-20, 1999, Cesena, Italy.
ISBN 0-9683689-6-4

1. Gravitation. 2. Electromagnetism. 3. Cosmology. I. Rudnicki, Konrad

QB981.G73 2001 530.11 C2001-900455-9

**Cover design by Dominic Turgeon**

# INTRODUCTION

Fritz Zwicky, the great 20th century astronomer, astrophysicist and theoretical physicist, also dealt with methodology of research, which is considered to be one of branches of the philosophy of science. Zwicky, unlike most philosophers working in this area, not only discussed methods used by others but applied his methodological ideas to a new practical approach in his highly successful scientific research. This approach helped him to discover new objects and new facts. His activity in the fields of the exact sciences and of philosophy *in* science formed an integral whole. He advocated taking all possible, even exotic hypotheses into consideration, and never adhering only to a single hypothesis. In his *Morphological Astronomy* he wrote the following words, which should be taken as a fundamental principle in all research:

> *If rain begins to fall on previously dry areas on the earth, the water on the ground will make its way from high levels to low levels in a variety of ways. Some of these ways will be more or less obvious, predetermined by pronounced mountain formations and valleys, while others will appear more or less at random. Whatever courses are being followed by the first waters, their existence will largely prejudice those chosen by later floods. A system of ruts will consequently be established which has a high degree of permanence. The water rushing to the sea will sift the earth in these ruts and leave the extended layers of earth outside essentially unexplored. Just as the rains open up the earth here and there, ideas unlock the doors to various aspects of life, fixing the attention of men on some aspects while*

> *partly or entirely ignoring others. Once man is in a rut he seems to have the urge to dig even deeper, and what often is most unfortunate, he does not take the excavated debris with him like the waters, but throws it over the edge, thus covering up the unexplored territory and making it impossible for him to see outside his rut. The mud he is throwing may even hit his neighbours in the eyes, intentionally or unintentionally and make it difficult for them to see anything at all.*

This volume, devoted to the problems of relativity, gravitation and related issues in physics, presents papers delivered and/or discussed during the conference "Redshifts and Gravitation in a Relativistic Universe" held in Cesena on September 17-20th 1999. In a way, this conference represents a response to Zwicky's method, outlined above. Its main aim was to serve as a forum for ideas and theories that go against the mainstream of science. Some of the theories are already cast in their final form; some are just rough ideas still undergoing development. Not all of them will prove correct, just as not all of the mainstream theories are wrong. Only reality is an absolute truth, while our theories have only approximate validity. The great German thinker Johann Wolfgang von Goethe wrote: *not distinguishing between reality and theory is like not distinguishing between a building and its scaffolding.* Theories are tools, not objects of scientific investigation, but indispensable tools. Only a wide variety of tools can enable us to carry out such a complicated task as scientific research.

In addition, a wide variety of observed phenomena have to be taken into consideration in a properly organized scientific investigation. Some phenomena which are seldom mentioned by others—such as quantization of redshifts—are discussed in this volume.

Some of the papers are presented here in more or less the same form in which they were delivered during the conference. Some were reworked more recently and take a final form different from the presentation. No minutes of the extensive discussion in the conference auditorium or the more lively discussions that continued during breaks and around dinner tables were recorded. In some cases the discussions are reflected in the final shape of the papers. Two of the papers included here were not presented as such during the conference, but their content was mentioned and taken under consideration during the debate. This volume therefore should not be regarded as a formal proceedings of the Cesena Conference, although it does fairly reflect the substance of the event.

In his contribution A.K.T. Assis proposes the principle of physical proportions, according to which all laws of physics can depend only on the ratio of known quantities of the same type. An alternative formulation is that all universal constants of physics (*G*, *c*, Planck's constant, Boltzmann's constant, *etc.*) must depend on cosmological or microscopic properties of the universe. There is a discussion of laws satisfying this principle and of other laws which do not follow it, implying that the corresponding theories must be incomplete. The author shows how to implement this principle by means of his theory of Relational Mechanics, as set out in the book of the same title (Apeiron, Montreal, 1999).

The paper presented by H. Broberg is based on the equivalence between gravitation and acceleration, initially suggested by Einstein. This introduces a new geometric approach to quantum gravity, the missing link to unification, extended to a discussion of energy flows in the vacuum as the key mechanism of the gravitational process. His ideas also relate to string theory in a scenario where the extra dimension, representing the "thickness of the line," can be allowed to exist from the Planck length up to the Hubble scale.

An alternative picture of the structure of galaxies is proposed in the paper by Marek Biesiada, Konrad Rudnicki and Jacek Syska. The authors discuss the possible explanation of dynamical properties of galaxies with the theory of dilatonic balls using six-dimensional space.

In the paper "Electromagnetism and Cosmology" by Edward Kapuścik a rather convincing argument is given that the correct unification of electromagnetism and gravity should start from some elementary and basic protofields which are neither electromagnetic or gravitational fields. The presently observed division of fundamental interactions into gravitational and electromagnic must be achieved by constructing composite fields from the protofields. In addition to the field equations, the gauge conditions also express physical laws and determine these composite fields. The last statement contradicts the point of view commonly adopted, which treats the gauge fields as auxiliary quantities.

Two papers by F. Selleri show that transformations of space and time between inertial systems exist which are almost empirically equivalent to the

Lorentz transformations. They contain a free parameter $e_1$, the coefficient of $x$ in the transformation of time. He shows that Michelson type experiments, aberration, occultation of Jupiter satellites, and radar ranging of planets are insensitive to the choice of $e_1$. An exception is represented by experiments in slowly accelerated frames, *e.g.*, those concerning the Sagnac effect. The best choice emerging from Selleri's work is where the parameter $e_1 = 0$, *i.e.*, a theory different from Special Relativity.

One of the goals of the Cesena conference was to find common ground among the dissidents beyond their certitude that some mainstream models are wrong. That proved surprisingly difficult, and the discussions showed why—we differed about which fundamental starting points were a valid basis for building models. Should model-building be driven my math or by physics? Are singularities allowed by reality? Can matter and energy be created or destroyed? Must the causality principle be respected? And so forth. One session on the last day of the conference was devoted to a discussion of these points, and we found that no unanimity existed about any of them. That led directly to the contribution by Van Flandern, "Physics has its Principles," which attempts to examine several such fundamental principles and show the consequences in each case of making a wrong assumption about its applicability or non-applicability. Whether or not this initial effort brings dissident views closer, it has certainly highlighted the points that must be resolved for any hope of a convergence of models and viewpoints in the future.

Many physicists point to the proper functioning of the International Atomic Time system (TAI) in order to support the postulate of Special Relativity Theory about the one-way isotropy of light velocity in every inertial system, which has never been demonstrated. Contrary to this view, Manaresi demonstrates that the proper functioning of the TAI system does not imply the one-way isotropy of light on the moving Earth. This means that the second postulate of Special Relativity still remains merely conventional.

Astronomical observations show that some fundamental cosmic properties come in discrete values. The ratio of observed properties, such as redshift or mass, for example, yields a ubiquitous factor of 1.23. In the paper by A. and J Rubčić and H. Arp in this volume the properties of fundamental particles such as leptons and quarks are examined. The surprising result is that they also obey this "quantization" rule. While there is no current explanation, these empirical results point to similar physical laws which extend from the smallest to the largest entities in the universe. This may lead to a physical understanding of redshift quantization.

A very straightforward paper by K. Rudnicki, W. Godłowski and A. Magdziarz presents a statistical elaboration of a very small sample of objects within the Iwanowska lines of galaxies and globular clusters. It shows that globular clusters, even located together with galaxies on the same lines, do not show redshift periodisation, whereas the galaxies do show the periodisation.

B. Bligh starts with some basic notions of thermodynamics to expose some of the errors made by cosmologists. Thermodynamic calculations require an *energy balance*. He then presents calculations on the Hot Big Bang Theory using data provided by cosmologists. The results are presented in a table and graphs which show that the Big Bang Theory cannot be true. Mr. Bligh also explains that thermodynamic calculations are most easily done with the aid of a *temperature-entropy diagram* for hydrogen, a method that is demonstrated in detail in his book *The Big Bang Exploded!*

Lastly, the paper by Cardone and Mignani deals with a problem that has been the subject of long-standing debate in the literature, namely the possibility of a breakdown of local Lorentz invariance (a subject revived in recent years, *e.g.*, by S. Coleman, S.L. Glashow and R. Jackiw). In their paper, Cardone and Mignani report the preliminary positive results of an experiment which seems to evidence a DC voltage across a conductor induced by the static magnetic field of a coil. This intriguing finding ought, of course, to be confirmed by further independent tests, aimed at excluding possible gravitational effects, among the other things.

*Konrad Rudnicki*

# Table of Contents

# Applications of the Principle of Physical Proportions to Gravitation

A.K.T. Assis[1]

We propose the principle of physical proportions, according to which all laws of physics may depend only on the ratio of quantities of the same type. We present examples of laws that satisfy this principle, and others that do not. These examples suggest that the theories leading to these laws must be incomplete.

*PACS numbers*: 01.55.+b (General physics), 01.70.+w (Philosophy of science).

*Keywords*: relative and absolute magnitudes, relational mechanics.

## The Principle of Physical Proportions

Newton, in his *Principia* (1687), introduced the concepts of absolute space, absolute time and absolute motion. Leibniz, Berkeley and Mach were against these concepts and proposed that only relative space, relative time and relative motion could be conceived and perceived by the senses. We agree with these authors and propose a generalization of their ideas through the principle of physical proportions, which can be stated as follows: "All laws of physics can depend only on the ratio of quantities of the same type." The meaning of this principle is illustrated by the examples below.

The law of the lever satisfies this principle. According to Archimedes two weights $P_1$ and $P_2$ at distances $d_1$ and $d_2$ from a fulcrum remain in horizontal static equilibrium only when $P_1/P_2 = d_2/d_1$. Only ratios of local weights and local distances are relevant here.

On the other hand, classical mechanics does not satisfy this principle. For instance, the acceleration of free fall near the surface of the earth is given by

[1] Instituto de Física "Gleb Wataghin" Universidade Estadual de Campinas, Unicamp, 13083-970 Campinas, São Paulo, Brasil. E-mail: assis@ifi.unicamp.br. Homepage: http://www.ifi.unicamp.br/~assis.

$$a = \frac{GM_e}{R_e^2} = \frac{4\pi}{3} G R_e \rho_e .$$

Here $G = 6.67 \times 10^{-11} Nm^2 / kg^2$ is the constant of gravitation, $M_e = 5.98 \times 10^{24} kg$ is the earth's mass, $R_e = 6.37 \times 10^6 m$ is its average radius and $\rho_e = 5.52 \times 10^3 kg / m^3$ its average mass density. This acceleration of free fall depends only on the mass (or density) of the earth, and not on the ratio of this mass (or density) to other masses (or densities) in the universe. Although the constant $G$ has the dimensions of acceleration divided by (distance times density), it is not dependent on other bodies in the universe, since it is a universal constant. This situation is in conflict with the principle of physical proportions.

Relational mechanics (Assis 1999, Section 8.1) has resolved this problem, as it is completely compatible with the principle of physical proportions. It is based on Weber's law for gravitation and electromagnetism, and on the principle of dynamical equilibrium: The sum of all forces of any nature (gravitational, electric, magnetic, elastic, nuclear, *etc.*) acting on any body is always zero in all frames of reference. As the sum of all forces is zero, only ratios of forces will be detectable or measurable. The system of units (MKSA, cgs, *etc.*) to be employed is not relevant. Moreover, the unit or dimension of the forces can be whatever we wish.

According to relational mechanics the acceleration of free fall is given by (Assis, 1999, Sections 8.5 and 9.2)

$$a = \frac{2}{\alpha} R_e H_o^2 \frac{\rho_e}{\rho_o} .$$

Here $H_o$ is Hubble's constant and $\rho_o$ is the average matter density of the distant universe. Moreover, $\alpha$ is a dimensionless number with value 6 if we work with a finite universe and integrate Weber's law for gravitation up to Hubble's radius $c / H_o$. If we work with Weber's law and an exponential decay in gravitation we can integrate up to infinity, and in this case $\alpha = 12$.

The important aspect of this result is that only a ratio of densities is important here. Doubling the earth's density while keeping the mass density of the distant universe unaltered is equivalent to keeping the earth's density unaltered while halving the mass density of the distant universe. In both cases the acceleration of free fall doubles compared to its present value of $9.8 m/s^2$.

Next we consider the figure of the earth.

## The Flattening of the Earth

Due to its diurnal rotation around the North-South axis the earth takes essentially the form of an ellipsoid of revolution. Its equatorial radius $R_>$ is greater than the polar radius $R_<$. According to classical mechanics the fractional change $f$ is given by (Assis, 1999, Section 3.3.2):

$$f \equiv \frac{R_> - R_<}{R_<} \approx \frac{15\omega^2}{16\pi\, G\rho_e} \approx 0.004 \, .$$

Here $\omega = 7.29 \times 10^{-5} rad/s$ is the angular rotation of the earth relative to an inertial frame of reference with a period of one day.

Several observations may be made considering this result, which is based on classical mechanics. In the first place the fractional change depends on the angular rotation of the earth relative to absolute space or to an inertial frame of reference. In principle, the distant universe composed of stars and galaxies can disappear without affecting $f$. If the earth remained stationary in an inertial frame of reference and the distant universe rotated around its North-South direction in the opposite direction compared with the previous situation, the earth would not be flattened. This is against Mach's point of view. Moreover, the fractional change depends only on the density of the earth, but not on the density of distant matter. If it were possible to double the average matter density of the distant universe without affecting the matter density of the earth, the previous result would not be affected. This shows that not only space and time, but also mass and matter density are absolute quantities in classical mechanics. All of these aspects are against the principle of physical proportions.

The flattening of the earth according to relational mechanics is given by (Assis, 1999, Sections 8.5 and 9.5.1)

$$f \equiv \frac{R_> - R_<}{R_<} \approx \frac{5\alpha}{8} \frac{\omega_{eU}^2}{H_o^2} \frac{\rho_o}{\rho_e} \, .$$

As there are many uncertainties concerning the precise value of Hubble's constant and the average matter density of the universe, it is not possible to give a precise value for the above ratio. But the order of magnitude is compatible with the observed value of 0.004. We can also utilize the fact that this is the observed value of *f*, and in this way (together with the known value of the angular rotation of the earth and its matter density) derive the value of $\alpha \rho_o / H_o^2$.

But what we want to emphasize here are the Machian aspects of this result. The first is that the angular rotation $\omega_{eU}$, which appears in relational mechanics, is the angular rotation of the earth relative to the distant universe (distant galaxies). It is no longer the angular rotation of the earth relative to free space. According to relational mechanics, there will be the same flattening of the earth no matter whether the earth rotates relative to an arbitrary reference frame while the distant universe remains stationary in this frame, or if the distant universe rotates while the earth remains stationary in this frame, provided the relative rotation between the earth and the distant universe is the same in both cases. The flattening of the earth can no longer be considered as a proof of the real or absolute rotation of the earth. The second aspect is that this flattening depends on the ratio of densities of the distant universe and of the earth. We can increase the flattening by decreasing the density of the earth or increasing the density of the distant universe. Only ratios of quantities are important here. Mass and matter density are not absolute in relational mechanics. The last aspect to be considered here is the ratio of the angular rotation of the earth and Hubble's constant. If we double the rotation of the earth relative to the distant universe, the flattening increases four times, as it is proportional to the square of the angular rotation of the earth. To say that the rotation of the earth has increased we must compare it with something else (for instance, with a clock). The same result should appear if the earth did not change its rate of rotation, but all other motions in the universe became slower by half. This means that Hubble's constant must somehow be like an average frequency of oscillation and/or rotation of the matter in the universe. If we decrease by a factor of two all of these frequencies (except the frequency of rotation of the earth relative to the distant universe), the present value of Hubble's constant must then be divided by 2, and the flattening increases by a factor of four, as in the previous situation. This happens in relational mechanics but not in classical mechanics.

## Applications to Other Situations

We now propose applications of this principle to other situations involving different physical concepts.

We first analyze electrostatics. Consider two charges $q_1$ and $q_2$ of the same sign repelling one another. We can keep them separated at a constant distance $d$ by applying an external force, for instance, placing a dielectric spring of elastic constant $k$ and relaxed length $\ell_o$ between them. By equating the coulombian force with the elastic force $k(d-\ell_o)$, we find that the fractional displacement $f$ of the spring is given by

$$f \equiv \frac{d-\ell_o}{\ell_o} = \frac{q_1 q_2}{4\pi\,\varepsilon_o\,d^2 \ell_o\,k}.$$

Here $\varepsilon_o = 8.85\times 10^{-12} C^2 s^2/kgm^3$ is called the vacuum permittivity. Doubling the value of the two charges increases $f$ four times. The fractional displacement should also increase four times according to the principle of dynamical equilibrium if $q_1$ and $q_2$ are kept unaltered but all other charges in the universe are halved (*i.e.*, the charges of all atoms and molecules of the spring, the earth and of all other bodies of the universe, excepting $q_1$ and $q_2$). However, this consequence is not implemented in present theories, indicating that they must be incomplete. The influence may be completely local (halving all the charges of the spring and distance galaxies changes only the elastic constant to $k/4$, without affecting $\varepsilon_o$), completely cosmological (halving all the charges of the spring and of all astronomical bodies does not change $k$, but does change the vacuum permittivity to $\varepsilon_o/4$), or a mixture of both effects (halving all the charges of the spring and of all astronomical bodies affects the elastic constant and the vacuum permittivity, their new values becoming $k/2$ and $\varepsilon_o/2$).

Suppose now we remove the spring, releasing the charges. They will then be accelerated in opposite directions. The value of the acceleration of $q_1$ relative to an inertial frame or to the universal frame of distant galaxies is given by

$$a_1 = \frac{q_1 q_2}{4\pi\,\varepsilon_o\,\mathrm{d}^2 m_1}.$$

This acceleration is increased four times when $q_1$ and $q_2$ are doubled. The same must happen when $q_1$ and $q_2$ are kept unaltered but all other charges in the universe are halved (that is, the charges of all atoms and molecules of distant galaxies, and the microscopic charges composing bodies 1 and 2 are all halved). Again the effect may be totally cosmological (affecting only the vacuum permittivity), totally local (affecting only the masses $m_1$ and $m_2$) or a mixture of both effects (affecting the vacuum permittivity and both masses).

One example of how the mass of a body may depend on its microscopic constituent charges has already been given (Assis, 1992). The Newtonian gravitational force between two bodies of masses $m_1$ and $m_2$ was derived as a residual electromagnetic force arising from the interaction between the neutral oscillating dipoles composing body 1 and the neutral oscillating dipoles belonging to body 2, where each dipole consisted of a negative charge oscillating around a positive one. The mass of each body was then found proportional to the number of oscillating dipoles composing it, and to $q^2/\varepsilon_o$, where $q$ represents the positive (or negative) charge of each neutral dipole.

Another situation is Ampère's force between electrical circuits carrying currents $I_1$ and $I_2$, proportional to $I_1I_2$. As the currents are proportional to the drift velocities of the electrons, we can increase the force four times, doubling these drift velocities. The consequences of this effect can be seen statically (an increase in the tension on a spring holding the two circuits at a constant distance) or dynamically (an increase in the acceleration of the two circuits when the spring is released). The same consequences must be found if we keep $I_1$ and $I_2$ unaltered, but make all other bodies in the universe move with half their present velocities. As modern theories do not implement this property, they must be incomplete.

Consider now the equation of state of an ideal gas, $PV = k_BNT$ ($P$ being the pressure, $V$ the volume, $k_B = 1.38 \times 10^{-23}$ $J/K$ Boltzmann's constant, $N$ the number of atoms and $T$ the temperature). This equation is not compatible with the principle of physical proportions. The equation of an ideal gas compatible with this principle should take the form $(p/p_o)(V/V_o) = a\,(N/N_o)(T/T_o)$, where "$a$" is a dimensionless number and $p_o$, $N_o$ and $T_o$ are local and/or cosmological pressures, the number of particles and temperature. When the theory leading to this new equation is found, it will be possible to relate Boltzmann's constant $k_B$ to the properties (such as pressure, density and temperature) of the local or cosmological environment. For instance, relational mechanics showed that the universal constant of gravitation $G$ is proportional to $H_o^2/\rho_o$. This shows that it is no longer a constant, but a function of the properties of the distant universe. An analogous situation must hold for Boltzmann's constant.

The same can be said of almost all relations in physics. The universal constants, such as the light velocity *in vacuo*, Planck's constant, *etc*. must all be functions of properties of the distant universe (macrocosm, holistic relations) or of local particles (microcosm, microscopic relations).

We hope this paper will motivate others to search for these relations in all branches of physics. Many new things will be learned in this process, and certainly many novel developments and deeper theories will come out of this endeavour.

## References

Assis, A. K. T. "Deriving gravitation from electromagnetism." *Canadian Journal of Physics*, Vol. 70 (1992), pp. 330-340.

Assis, A. K. T., *Relational Mechanics*. (Montreal, Apeiron, 1999), ISBN: 0-9683689-2-1.

# The Geometry of Acceleration in Space-Time

## Application to the Gravitational Field and Particles

Henrik Broberg[2]

## Introduction

In a paper presented June 1997, at a conference in Athens (Ref.1), the author developed a holistic view of the Universe and its components, all joined together in a common geometry in four-dimensional space-time, applied to the Universe as a whole, as well as to its constituent components—the particles.

The ideas documented here, which form a continuation of the Athens paper, were introduced in a first draft form February 1999 at a lecture to the Indian Institute of Technology in Kharagpur and later in September the same year at the Cesena conference. They initially concern a new approach to quantum gravity, the missing link to unification, but also extend to a discussion of energy flows in the vacuum as the mechanism of the gravitational process. The ideas introduced here are also related to string theory, although in a transformation scenario where differentials of any size are allowed, and therefore an extra dimension, representing the "thickness of the line," can be allowed from the Planck length up to the Hubble scale.

## Specific features of the presentation

The equivalence between gravitation and acceleration suggested by Einstein is a cornerstone for the work presented here.

All distances are defined as space-time objects, in accordance with General Relativity. Mathematically, the differentials used are valid space-time objects, without any need for limits of size, except for the changing conditions set by the transfers into the complex domain of numbers. The $\Rightarrow$ sign is used frequently for "can be developed to...."

[2] henrik.broberg@telenor.com

## The Concept of Time

Before Einstein introduced the definition of time as a geometric property in four-dimensional space-time, there was no consistent definition of time. Nobody could even say what time was, beyond an effort to measure rate of change in terms of one process or another, such as the ticks of a clock or the rotation of Earth around the sun, *etc*. As long as the light-velocity $c$ is used to define time, it will, of course, not be possible to introduce variations in the quantitative number given to $c$, which might equally as well be given the number 1 as any other number. For simplicity, $c = 1$ is frequently used in physics.

However, the physical group-velocity of electromagnetic waves, or physical light, does (as we know) show variations depending on the medium, which might even be the case for the vacuum under certain circumstances.

In the following, Einstein's definition of time is used, *i.e.*, time is defined as a distance in four-dimensional space-time.

In the four dimensions, the time-like scenario is "frozen" into a sculpture without causality, while it is in the three dimensions that causality, cause and effect are given a meaning. Still, we need the four dimensions to define the time which we experience in the three dimensions.

## The concept of acceleration

The concept of acceleration in mechanics is related to the application of a force to an object. As long as no counter-forces are present, the object will be accelerated, meaning it is forced to change its velocity in comparison with its earlier state of rest. This will also mean that the object is transferred into another time system. In order to deepen our understanding of the concept of gravitation, we need therefore to give a meaning to acceleration in a central force field as a space-time concept.

The change of velocity of an object—say a mass-point for simplicity—can be regarded as a change from one frame, *i.e.* the one where the particle is at rest, into another frame, which has a certain velocity in relation to the first one, where the particle will accommodate itself in a new state of rest, at least momentarily, after having undergone the process we call "acceleration."

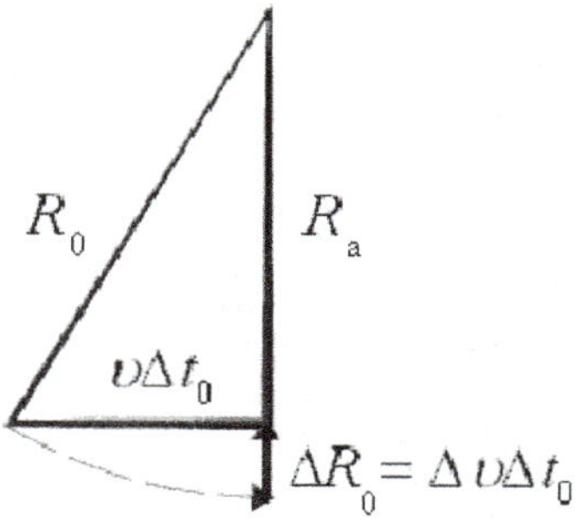


**Figure 1**. Transfer from Frame O to Frame A.

We first introduce Frame O as the frame where the particle is initially "at rest." Frame A is the accelerated frame, into which the particle will be transferred by acquiring a certain velocity $\upsilon$ with respect to Frame O.

With reference to Figure 1, we introduce: $R_0 = c\Delta t_0$ and $R_a = c\Delta t_a$.

Due to the "boost" which brings the particle into Frame A, the space-time becomes shortened by $\Delta R_0$ to $R_a$, where $\Delta R_0 = \Delta\upsilon_0 \cdot \Delta t_0$. When Frame O is transferred in this way into Frame A, $R_o$ will line up with $R_a$ and, most important, it will be shortened to $R_a$ by a displacement in the space-time geometry. This latter effect will be associated with a velocity differential $\Delta\upsilon$ acting during the time $\Delta t_0$. Hence $R_0 - R_a = \Delta\upsilon \cdot \Delta t_o$.

This leads to the relation

$$\frac{R_o}{R_a} = \frac{1}{1 - \frac{\Delta\upsilon}{c}}.$$

We write the relation between the times, mapped as $R_0$ and $R_a$, as

$$\gamma = \frac{1}{1 - \frac{\Delta\upsilon}{c}},$$

and therefore also

$$\gamma - 1 = \frac{\frac{\Delta\upsilon}{c}}{1 - \frac{\Delta\upsilon}{c}}.$$

The energy requirement for the acceleration process can now be expressed by $E = F \cdot S$, where $F$ is a force and $S$ a distance.

We introduce the force as $F = am$, and $S = R_0$. The acceleration is given by $a = \Delta\upsilon / \Delta t_0$.

Hence, we get the energy as

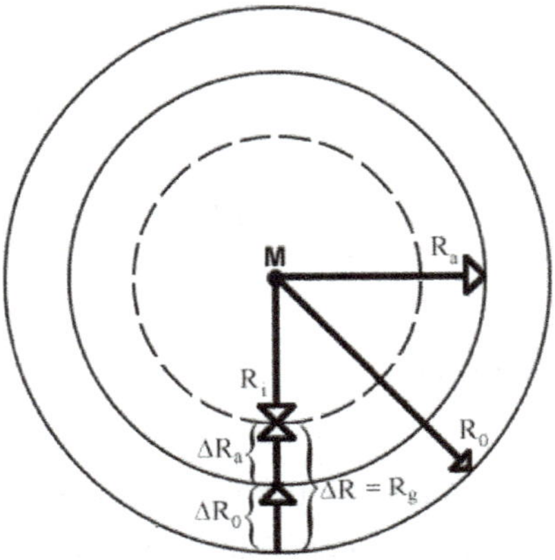


**Figure 2.** The space around a central mass.

$$E = \frac{\Delta \upsilon}{\Delta t_0} \cdot m_0 \gamma \cdot R_0 \equiv \frac{\Delta \upsilon}{\Delta t_0} \cdot m_0 \cdot \frac{1}{1 - \frac{\Delta \upsilon}{c}} \cdot c\Delta t_0 \equiv m_0 c^2 \cdot \frac{\frac{\Delta \upsilon}{c}}{1 - \frac{\Delta \upsilon}{c}} \equiv m_0 c^2 (\gamma - 1),$$

which is equivalent to the well known form for kinetic energy from Special Relativity (SR), although here the energy is also given in geometric terms by the differential displacement of the frame of the particle as it is transferred from one time-system into another in four-dimensional space-time. In the space-time picture of the acceleration process, the energy-force relation is apparently inherent in the geometry, as if the force caused by the acceleration at the radius vector distance from a centre (here represented by the top of the triangle in Figure 1) were transferred all along the radius vector. Hence, the integration is inherent in the picture locally in the neighbourhood of the point.

## The field of gravitation

The above-described geometric concept of acceleration in space-time is also applicable to a central force field, in which case the time-differentials will apply to radius vectors from the centre, *e.g.*, a mass-point. Here it will be used for the description of the gravitational field, with reference to Figures 3-5.

The gravitational radius ( $R_g$ ) of a mass ( $M$ ) is given with reference to the Schwarzschild solution, and has the same value as the radius to the event horizon of a black hole with the energy content of the mass. The concept of a mass-point should therefore be understood as equivalent to a black hole. We have $R_g = 2GM/c^2$, where $G$ is Newton's constant. Due to the proportionality between the gravitational radius and the mass, the latter can also be treated as a system of small mass-points, represented by the sums of their masses and the sum of their gravitational radii in respect of the gravitational centre.

Initially, the space around a mass-point is described using Figure 2, without any restriction to the directions of radius vectors. In order to describe the geometry of the gravitational field, we define

$$\begin{cases} R_0 = c\Delta t_0 \\ R_a = c\Delta t_a \\ R_i = c\Delta t_i \end{cases},$$

where $R_0$ is the distance from the point marked $M$ in the case where there is no mass at the point, while $R_a$ is the corresponding distance when there is a mass $M \neq 0$ at the point. Hence, due to the influence of the mass, the distance to the point has shrunk to $R_a$. In the same way, the distance $R_a$ would shrink to $R_i$ when $M$ is introduced. Therefore, also:

$$R_0 - R_a = \Delta R_0 ,$$

Multiplying both sides of the above equality by $(R_0 + R_a)$ gives:

$$(R_0 + R_a)(R_0 - R_a) = (R_0 + R_a) \cdot \Delta R_0$$

which can be developed to:

$$R_a^2 = R_0^2 - (R_0 + R_a) \cdot \Delta R_0$$

and further to

$$\frac{R_a^2}{R_0^2} = 1 - \frac{(R_0 + R_a) \cdot \Delta R_0}{R_0}$$

We associate the displacement of the vacuum space around $M$ with a velocity differential $\Delta \upsilon_0$ defined by

$$\Delta R_0 = \Delta \upsilon_0 \cdot \Delta t_0 .$$

Substituting with the time-equivalents of $R_0$ and $R_a$ now leads to

$$\frac{\Delta t_a}{\Delta t_0} = \sqrt{1 - \frac{(\Delta t_0 + \Delta t_a) \cdot \Delta \upsilon_0}{c \Delta t_0}} .$$

With the (thus far, *ad hoc*) substitution of $R_g = \Delta\upsilon(\Delta t_0 + \Delta t_a)$, we get

$$\frac{\Delta t_a}{\Delta t_0} = \sqrt{1 - \frac{R_g}{R_0}},$$

which is identical with the time dilation in a gravitational field, in accordance with the Schwarzschild solution to the vacuum field equations of General Relativity. We therefore adopt the hypothesis that $\Delta R = R_g(M)$, or for simplicity $\Delta R = R_g$.

We can now also develop the contraction of $R_0$ in the field by multiplying the above formula by $c$ in the nominator and denominator:

$$\Delta R_0 = R_0 - R_a \Rightarrow R_0\left(1 - \sqrt{1 - \frac{R_g}{R_0}}\right).$$

This is the gravitational length contraction, also in agreement with General Relativity, which further strengthens our hypothesis that in the gravitational field, $\Delta R = R_g$, to be further proved in the following chapter.

## The geometry of the field

We associate the displacement of the vacuum space around $M$ with velocity differentials $\Delta\upsilon_0$ at $\Delta R_0$, $\Delta\upsilon_a$ at $\Delta R_a$ and $\Delta\upsilon_i$ at $\Delta R_i$. Hence, we have

$$\begin{cases} \Delta R_0 = \Delta\upsilon_0 \cdot \Delta t_0 \\ \Delta R_a = \Delta\upsilon_a \cdot \Delta t_a \\ \Delta R_i = \Delta\upsilon_i \cdot \Delta t_i \end{cases}.$$

The geometry of the field can be illustrated by Figure 3.

The cycle begins with the transfer of an object from its state of rest in the time-system of $R_0$ (Frame O) into the time-system of $R_a$ (Frame A), while the vacuum is displaced $\Delta R_0 = \Delta\upsilon_0 \cdot \Delta t_0$. To ensure compliance with the definition of acceleration in space-time illustrated in Figure 1, a velocity parameter $\upsilon_g$ is introduced in Figure 3. In this case, and assuming $\Delta R = R_g$, we find from the geometry that $\upsilon_g{}^2 = \left(R_g / R_0\right)c^2$, a relation which satisfies the energy relation for an object with mass $m_0$, acted on by the force $F = -m_0\gamma_g \cdot a$ over the distance $R_0$.

$$-F \cdot R_0 = m_0 c^2(\gamma_g - 1).$$

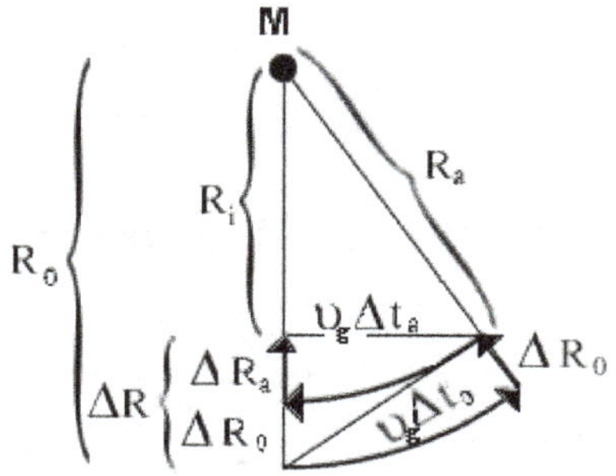


**Figure 3.** The space-time geometry of the gravitational field.

Thus $\upsilon_g$ becomes equal to "the classical free fall velocity" (referring to a fall from "infinite distance") in the gravitational field.

To prevent the object from "falling" towards $M$, there must be a counter action, such that $\Delta R_0$ is cancelled out during the time differential $\Delta t_0$, as in the example of the centrifugal force, given in the following chapter. We will now evaluate the relations between the velocity differentials in the gravitational vacuum field. We therefore have

$$\begin{cases} \Delta\upsilon_0 = c \cdot \dfrac{R_0 - R_a}{R_0} \\ \Delta\upsilon_a = c \cdot \dfrac{R_a - R_i}{R_a} \end{cases},$$

giving

$$\frac{\Delta\upsilon_0}{\Delta\upsilon_a} = \frac{R_0 - R_a}{R_a - R_i} \cdot \frac{R_a}{R_0} \Rightarrow \frac{R_0 R_a - R_a^2}{R_0 R_a - R_0 R_i}.$$

From the geometry of Figure 3, we have the relation

$$\frac{R_a}{R_0} = \frac{R_i}{R_a}, \text{ or } R_a^2 = R_0 R_i,$$

which leads to $\Delta\upsilon_0 = \Delta\upsilon_a$. This result confirms the quantum aspect of the region between $R_0$ and $R_i$.

We will now evaluate the space-time distance in a round-trip from the mass-centre (mass-point) at $M$ to $R_0$ and back. We write

$$R_0 - \{R_i + \Delta\upsilon_0(\Delta t_0 + \Delta t_a)\} = 0,$$

yielding

$$\Delta \upsilon_0 (\Delta t_0 + \Delta t_a) = R_0 - R_i \Rightarrow R_g ,$$

which gives the physical rationale behind our earlier proposition that $\Delta R = R_g$.

We have already seen that this value should be equal to the gravitational radius associated with the mass *M*, in order to satisfy the relations of the Schwarzschild metric. Hence, the step from $R_0$ to $R_i$ should be equal to the quantum $R_g$ in the space surrounding *M* at all distances from the point. In the present context, this will mean that the vacuum space within the distance $R_g$ from the point at *M* is absorbed during each round-trip cycle.

The acceleration in the gravitational field becomes (using the space-time relations of the radii)

$$\frac{\Delta \upsilon_0}{\Delta t_0} = \frac{c^2 R_g}{R_0 (R_0 + R_a)} \approx \frac{GM}{R^2} ,$$

where the right hand side is recognised as the "classical" Newtonian gravity acceleration, valid for $R_0 \succ\succ R_g$.

At the gravitational radius, $R_a \to 0$ and $\Delta \upsilon_0 / \Delta t_0 \to 2GM/R_0^2$, which is also in agreement with the Schwarzschild solution for the pattern a light beam would follow at the event horizon.

The above calculated roundtrip from $R_0$ to *M* and back can also be regarded as a vacuum flow leaving the tip of $R_0$ with the constant velocity $\Delta \upsilon_0$ simultaneously with a light-beam of velocity *c*, which turns around at *M* and meets the flow at $R_i$ after having travelled the distance $R_g = \Delta \upsilon_0 \cdot \Delta t_0 + \Delta \upsilon_i \Delta t_i$. Comparing this with the earlier expression for the same distance, $R_g = \Delta \upsilon_0 \cdot \Delta t_0 + \Delta \upsilon_0 \Delta t_a$, we find the relation

$$\Delta \upsilon_i = \Delta \upsilon_0 \cdot \frac{\Delta t_a}{\Delta t_i} \equiv \Delta \upsilon_0 \frac{R_0}{R_a} .$$

This constitutes the "hand-shake" between two contiguous quantum regions in the vacuum space, while the $R_i$ of one becomes the $R_0$ of the next in the sequence towards *M*. With reference to Figure 3, $\upsilon_g \Delta t_a$ will at the same time rotate from its horizontal position and take the place of $\upsilon_g \Delta t_0$, which now becomes $\upsilon_g' \Delta t_0'$, with the value $\sqrt{R_i R_g}$ and perpendicular to the new $R_a' = \sqrt{R_i (R_i - R_g)}$.

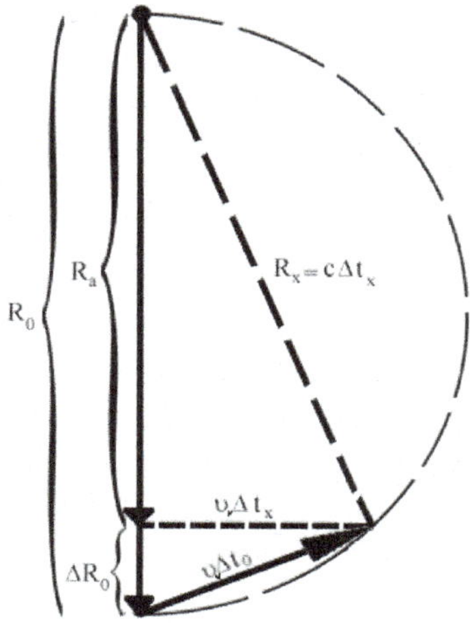


**Figure 4.** The space-time relations of a particle under rotation around a central point.

The velocity parameter $\upsilon_g$ changes to $\upsilon_g' = \upsilon_g \cdot \sqrt{R_0/R_i}$ (if we remember that $R_i = R_0 - R_g$). The angle ($\alpha$) between $R_0$ and $R_a$ increases such that $\sin\alpha' = \sin\alpha \cdot \sqrt{R_0/R_0 - R_g}$. This process will gradually enlarge the angle until it finally becomes 90 degrees when $R_0$ approaches the gravitational radius, *i.e.*, corresponding the situation when light follows the displacing space in a circle around the singular point at the event horizon.

## The centrifugal force and the gravity field

In the case of a planet orbiting the sun, or any other arrangement of a body orbiting a centre of mass, the difference between the centrifugal acceleration in the direction from the mass-centre and the gravitational acceleration towards the mass-centre will cause the resulting force on the body.

In the case of the centrifugal force, the displacement of the vacuum frame toward the mass-centre will cancel out the displacement of the position of a moving object in the radial direction, thus maintaining an equal distance to the mass-centre M in a circular orbit.

The geometric relations of the centrifugal force in Figure 4 are limited to the domain between $R_0$ and $R_a$, defined by $\Delta R_0$, where we now also need to include the interim level $\Delta R_x = R_x - R_a$ (Ref. to Figure 5). The velocity of rotation is $\upsilon_r$.

Simultaneously with the gravitational displacement of the vacuum from $R_0$ to $R_a$ we have an equal displacement in the opposite direction due to the centrifugal action. This latter displacement will take place in two steps, transferring from $R_a$ to $R_x$ and from there to $R_0$, as shown in the picture. The

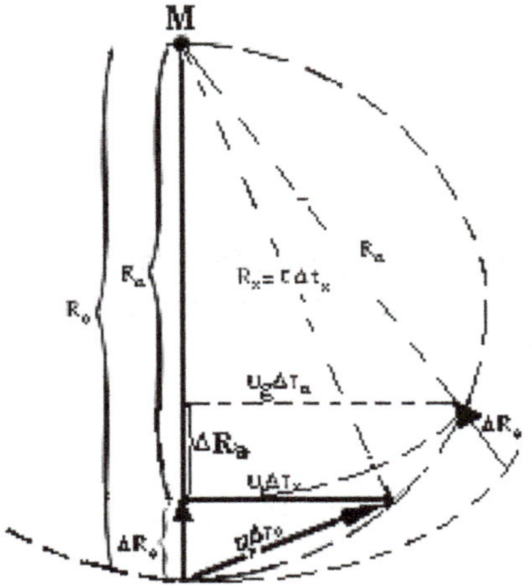


**Figure 5** Gravitation in balance with the centrifugal effect.

transfers from $R_0$ to $R_x$ and onwards to $R_a$ follow the same procedure as described before in respect of accelerating systems, and correspond in value to making the Lorentz transformation twice.

From Figure 4, the displacement of the radius from the centre of the rotation (top of the large triangle) is

$$\Delta R_0 = \sqrt{(\upsilon_r \Delta t_0)^2 - (\upsilon_r \Delta t_x)^2} \Rightarrow \frac{\upsilon}{c}\sqrt{R_0^2 - R_x^2} \Rightarrow \frac{\upsilon_r{}^2 \Delta t_0}{c}.$$

Figure 4 and Figure 3, when partly superimposed, give the picture of Figure 5 showing how the gravitational effect on the vacuum cancels out the centrifugal effect.

The acceleration towards the centre is

$$a = \frac{\Delta R_0}{(\Delta t_0)^2} \Rightarrow \frac{\upsilon_r{}^2}{R_0},$$

With $R_0 = c \cdot \Delta t_0$, we also find that

$$\frac{\Delta R_0}{R_0} = \frac{\upsilon_r{}^2}{c^2}.$$

For the gravitational field, we already have the relation

$$\frac{\Delta R_0}{R_0} = 1 - \sqrt{1 - \frac{R_g}{R_0}}.$$

In a situation of equilibrium, the two displacements will equal out, and we have

$$\frac{1}{1-\frac{\upsilon_r^{\,2}}{c^2}} = \frac{1}{\sqrt{1-\frac{R_g}{R_0}}},$$

giving the relation between the relativistic factor and the factor introduced here for the time relations in the gravitational field: $\gamma_r^{\,2} = \gamma_g$.

When $R_0 \to R_g$ we get that $\upsilon_r \to c$ and $\Delta\upsilon \to c$, and the acceleration in the field becomes

$$a \Rightarrow \frac{c^2}{R_0} = \frac{2GM}{R_0^2},$$

which would be the case at the event-horizon of a black hole.

When $R_0$ is much larger than $R_g$, we recognise again the "classical" expression for the gravity field, resulting from Newton's law,

$$a = \frac{GM}{R^2}.$$

## Quantum Concepts

From the preceding discussion, we have identified a quantizing of the radius vector from a mass-point by the differential:

$$\Delta R = R_0\left(1-\sqrt{1-\frac{R_g}{R_0}}\right)$$

In case, that $R_g = R_G \geq R_0$ the expression can be rewritten as

$$\Delta R = R_0\left(1-i\sqrt{\frac{R_G}{R_0}-1}\right)$$

We note that the surface-differential of the ring between $R_0$ and $R_i = R_0 - \Delta R$ becomes $\Delta\Phi = \pi R_0 R_g$ whether it is calculated as $\pi(R_0^2 - R_i^2)$ or $\pi\Delta R\overline{\Delta R}$, and independently of whether $R_0 \geq R_i$. The expression $\pi\Delta R\overline{\Delta R}$ has the same form as the surface calculated from the Schroedinger wave-function, proportional to a probability density, and it can be shown that they are identical. However, before going further it must be realised that the introduction of the complex form for $\Delta R$ is the same as entering inside the event-horizon of the system under consideration, *e.g.*, the Universe itself, a particle or a quantum.

In the complex form, a wave function can be constructed from $\Delta R$ as follows.

We introduce the angle $\varphi$ in the complex plane, such that

$$\begin{cases} \sin\varphi = \sqrt{\dfrac{R_0}{R_G}} \\ \cos\varphi = \sqrt{1 - \dfrac{R_0}{R_G}} \end{cases}.$$

The angle $\varphi$ so defined becomes the angle in Figure 3 (at M in the top) modified such that $R_G$ replaces $R_0$ and $R_0$ replaces $R_g = \Delta\upsilon \cdot (\Delta t_o + \Delta t_i)$. In this way, $R_0$ becomes part of the gravitational radius $R_G$ of the surrounding "quantum universe," and we have from the figure that $R_G - R_G \cos^2\alpha = R_0$, or $\sin^2\alpha = R_0 / R_G$.

This makes the following two expressions identical:

$$\begin{cases} \Psi = \sqrt{R_0 R_G}\left(\sin\varphi - i\cos\varphi\right) \\ \Delta R = R_0\left(1 - i\sqrt{\dfrac{R_G}{R_0} - 1}\right) \end{cases}.$$

We also have, therefore, the complex conjugate function

$$\overline{\Psi} = \sqrt{R_0 R_G}\left(\sin\varphi + i\cos\varphi\right)$$

## The displacement velocity

In the complex framework, we can now develop the expression for the velocity differential $\Delta\upsilon$ pertinent to the vacuum displacement as acceleration in general, or more particularly in the case of gravitation, from the preceding:

$$\Delta\upsilon = c\cdot\left(1-\sqrt{1-\frac{R_G}{R_0}}\right)$$
$$\equiv c\cdot\left(1-i\sqrt{\frac{R_G}{R_0}}\cdot\sqrt{1-\frac{R_0}{R_G}}\right).$$

By substituting $\cos\varphi$ and $\sin\varphi$ in the above given expressions we get

$$\Delta\upsilon = c\cdot(1-i\cot\varphi).$$

This result is interesting in two important respects:

1. it has the constant velocity $c$ as the real term, and
2. it allows quantum effects to be dispersed over any area within the singular (quantum) domain, without limitations on velocity.

From the first point, it would therefore appear that light, for example, would transfer with the real-term velocity $c$ of the vacuum displacement, resulting from a process of acceleration. This is indeed the case, because light is generated when electrons shift between positions where they have different energy levels in the electrical fields inside the atoms. The electrons in the atom therefore constitute the quantum system in which the displacement (light-) wave is generated. Outside the atomic quantum, the wave proceeds in the quantum system of the Universe.

The second point would take care of contradictions between Relativity and Quantum Theories based on the assumed limitation to the velocity $c$ for the transmission of quantum effects, *i.e.*, as expressed in the Bell inequalities and the efforts to test them, *e.g.*, the experiments of A. Aspect, which have come out in favour of quantum correlations without the time-confinement required by special relativity.

## The electromagnetic wave

According to the above demonstration, the electromagnetic wave can be assumed to have a complex displacement velocity,

$$\Delta\upsilon = c\cdot(1-i\cot\varphi).$$

The real term $c$ is the invariant property used in the Lorentz transformations and in the Einstein definitions of time and space, while the imaginary term is associated with the concepts of quantum physics, as described in the following.

The earlier developed wave-function can be rewritten as follows:

$$\Psi_1 = R_G\left(\sin^2\varphi - i\sin\varphi\cos\varphi\right).$$

We introduce a variation to this function, $\Psi_2$, by shifting the angle to $\varphi + \pi/2$. Hence

$$\Psi_2 = R_G\left(\cos^2\varphi + i\cos\varphi\sin\varphi\right).$$

This gives the following relations:

$$\begin{aligned} &\Psi_1 + \Psi_2 = R_G \\ &\Psi_1\overline{\Psi_1} = R_0 R_G \equiv R_G^2 \sin^2\varphi \\ &\Psi_2\overline{\Psi_2} = R_G^2 - R_0 R_G \equiv R_G^2\cos^2\varphi \end{aligned}.$$

From the two latter follows

$$\Psi_1\overline{\Psi_1} + \Psi_2\overline{\Psi_2} = R_G^2.$$

Hence, we have defined a quantum system composed of two components whose surfaces and radii sum up to those of the system they compose together, independently of the angle $\varphi$. The angle could, for example, have the time dependent value $\omega \cdot t$, while the components obey the conservation laws (Ref.1) of their own "universe." The two components would be the electric and magnetic fields of an electromagnetic wave, which propagates with $c$ as its real velocity component for the displacement.

In the quantum world, the parameter $R_G$ may not necessarily be the gravitational radius in the classical sense, but could also be the radius applicable to any autonomous quantum system, as originally proposed in Ref. 1 and briefly described below.

## The quantum system

Based on the preceding analysis, the concept of a quantum system can be introduced, here defined as a generalized Schwarzschild region limited by its event-horizon:

1. It has a characteristic radius, which can be understood as the radius of its event horizon, or in a Riemannian sense, as the curvature of its space.
2. The surface across the event horizon—or in the concept of string theory, the cross-section of the string—is proportional to the mass of the system. (The constant of proportionality is here given as $A$).
3. The quantum system will contain at least one quantum, itself, or it may be composed of series of sub-quanta, which are defined in terms of their characteristic radii and surfaces, which sum up to those of the main system.

These statements can be mathematically expressed as follows:

I. $R_S = \dfrac{2g_S M_S}{c^2}$, $g_S$ is defined as Newton's "constant"; $g_U = G$.

II. $A = \dfrac{\pi \cdot R_S^2}{M_S}$, where $A$ is defined as a true constant of nature

III $$\begin{cases} R_S = \sum_v r_v \\ \Phi_S = \sum \phi_s \end{cases}$$

The above algebraic expressions can be realised, for example, by a geometry where the little radii $r_v$ sum up to the diameter $2R_S$ of a sphere, while their projections on surface segments $\phi_v$ sum up to the surface of the sphere. This holds also when the little radii are randomly reorganized, *i.e.*, they might equally well be considered as placed at one of the poles as somewhere in-between.

Combining I and II gives:

$$A = \frac{2\pi \cdot R_S g_S}{c^2},$$

and

$$G = \sqrt{\frac{Ac^2}{4\pi \cdot M_S}}.$$

Hence, if $A$ is a true constant of nature, $g_S$ will vary, depending on the mass (or energy) content of the system. To test this hypothesis, we relate $g_S$ to the system of a nucleon, while $A$ is calculated from Newton's and Hubble's constants, applicable to the present state of the Universe.

In line with above, assuming that, $R_U = c/H$, we get

$$A = \frac{2\pi \cdot G}{Hc} \approx 0.7 \quad \left( {m^2}/{Kg} \right),$$

and

$$M_U = \frac{\pi \cdot R_U^2}{A} \approx 10^{54} \text{ Kg},$$

which is the compound mass of $10^{11}$ galaxies, each composed of $10^{11}$ stars of an average mass equal to that of the Sun, about what has been estimated from astronomical observation as the mass of the "Universe."

The radius corresponding to a particle quantum, or the generalized Schwarzschild domain of the nucleon mass, would be

$$R_{nucl.} = \sqrt{\frac{AM_{nucl.}}{\pi}} \approx 10^{-14} \text{m},$$

which agrees with observation. In this case the gravitational constant would be transformed to

$$G_{nucl.} = \sqrt{\frac{Ac^4}{4\pi M_{nucl.}}} \approx 10^{40} \cdot G_{Newton},$$

which is in agreement with the known relation between the nuclear and gravitational forces.

The concept of the quantum system and its dependence on the constant $A$ seems, therefore, to apply to the microcosm of the elementary particles and the quarks, as well as to the Universe and beyond.

In line with above, the Universe can be modelled as a large particle quantum system, containing a sequence of sub-singularities with different "gravitational" forces, stronger as the masses become smaller. This would mean that the value of $G_{Newton}$ would be valid for the force field that binds together all the particles of the Universal quantum system at the present time, while other stronger forces will exist inside linked up with local quantum systems, and maybe even weaker forces emanating from larger structures outside our "Universe."

## The particle embedded in the geometry of the Universe

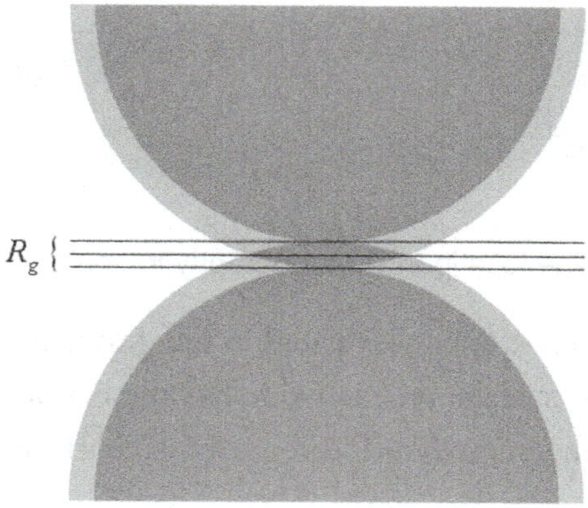


**Figure 6.** The local particle space-time attached to the Universal space from two sides.

A particle is represented by a quantum system, which is embedded in the system of the Universe. The cross-section of the particle can be calculated, assuming it to be the difference between the Universal cross-section with or without the particle mass,

$$\begin{cases} R_+ = \sqrt{\dfrac{AM_U}{\pi}} \\ R_- = \sqrt{\dfrac{A(M_U - M_P)}{\pi}} \end{cases}.$$

The particle cross-section becomes

$$\Phi_P = \pi \cdot (R_+^2 - R_-^2) \Rightarrow AM_P = \pi \cdot R_g \cdot R_U .$$

In Figure 6, the surrounding Universe is pulled in from two opposite sides to the level of the gravitational radius, forming a dish-like shape at the two attached polar regions, as illustrated in Figure 7, which shows the overlap of the curved space from the two sides cut out from the Universal space-time, by analogy with two bubbles of soap giving up some of their original surfaces when glued together; it may be interpreted as a hole, or a sink for vacuum energy, which could trigger the gravitational effect.

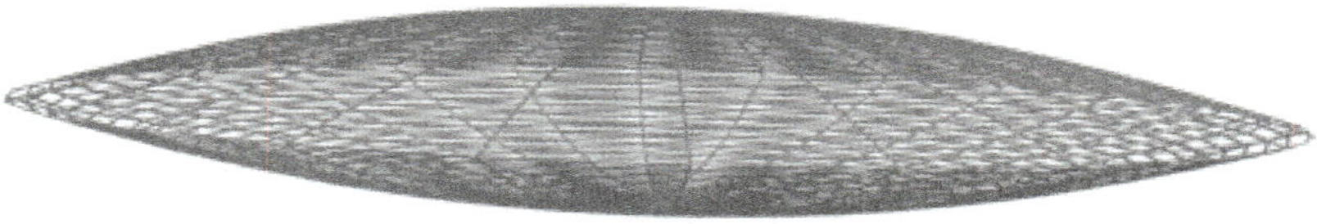

**Figure 7**. The local particle topography in space–time.

The above object has a thickness equal to the gravitational radius of the particle mass, while the surface on each side has the curvature radius of the Universe. The area of the curved surface on each side of the dish is equal to $\Phi_p$ from above.

At this point, the mental loop is almost closed back to the introduction of the gravitational field. On the way, the point-like particle has undergone the transformation to a dish-like object in space-time.

## The vacuum flow

To complete the loop also in a physical sense, we will finally discuss the vacuum displacement in terms of vacuum flows. This is consistent with the foundation of General Relativity, which derives its mathematical formalism largely from hydrodynamics, and is in essence a tensor based theory applicable to flows of "space," frozen into four-dimensional space-time. The flows themselves need to be reviewed also in a three dimensional context where causality, cause and effect can be given a meaning. The vacuum flow will need to be given a content, which can only be done in terms of energy, or energy equivalents.

In the two-dimensions of a plane through the mass-point, the surface covered during the time $\Delta t_0$ by a flow of displacement velocity $\Delta \upsilon_0$ perpendicular to the periphery, will be $\Delta t_0 \cdot \Delta \upsilon_0 \cdot 2\pi R_0$. The surface covered by the flow from $R_0$ to $R_a$ can be calculated as the average of flows over the peripheries of rings at $R_0$ and $R_a$, thus

$$\Delta\Phi_0 = \frac{2\pi \cdot \Delta t_0 \cdot \Delta \upsilon_0 \cdot (R_0 + R_a)}{2} \Rightarrow \frac{2\pi R_0}{2} \cdot \Delta \upsilon_0 (\Delta t_0 + \Delta t_a).$$

We recognise that the latter term is equal to $R_g$, which gives the surface element in the form

$$\Delta\Phi_0 = \pi R_0 R_g .$$

The same result is achieved by calculating the difference of the surfaces of rings with radii $R_0$ and $R_a$

$$\Delta\Phi_0 = \pi\left(R_0^2 - R_a^2\right) \Rightarrow \pi R_0 R_g .$$

Therefore, the flow completely covers the surfaces between consecutive rings.

A vacuum flow velocity, $\upsilon_f$, can now be introduced by calculating the surface covered in two-ways:

$$2\pi R_0 \cdot \upsilon_f \cdot \Delta t_0 = \pi R_0 R_g .$$

Hence, we get

$$\upsilon_f = c \cdot \frac{\frac{1}{2} R_g}{R_0} .$$

Therefore, during each time unit $(\Delta t_0)$ the flow covers the invariant distance $\upsilon_f \Delta t_0$, which becomes

$$\delta = \tfrac{1}{2} R_g .$$

Another invariant property is the flow-surface-rate

$$\frac{\Delta\Phi_0}{\Delta t_0} = \pi c R_g .$$

Substitution with the constant $A$ introduced earlier gives

$$\frac{\Delta\Phi}{\Delta t_0} = A H_0 M ,$$

where, as before, $A$ has the dimension $m^2/Kg$, $H_0$ is Hubble's constant (understood as an inverse Universal time parameter) and $M$ is the mass at the origin of the gravitational field.

The vacuum energy transported towards the singular mass-point per time unit ($\Delta t_0$) should be independent of the distance to the point, as long as no other sources or sinks for energy are encountered.

We have introduced the proportionality constant $A$ for the surface-to-mass ratio of the vacuum flow ($A/c^2$ for the contained energy) and $\rho$ for the (mass-) density of the vacuum. The following relations can be set up:

For the surface between consecutive rings at distances $R_0$ and $R_a$ from the central mass

$$A \cdot \Delta M = \pi\left(R_0^2 - R_a^2\right) \Rightarrow \pi R_0 R_g .$$

The amount of vacuum energy, in mass terms, streaming through any spherical shell central to the mass-point, during the time $\Delta t$, becomes

$$\Delta M = 4\pi R^2 \rho \upsilon_f \Delta t .$$

Hence, we have the equation system

$$\begin{cases} A \cdot \dfrac{\Delta M}{\Delta t} = \pi c R_g \\ \dfrac{\Delta M}{\Delta t} = \upsilon_f \cdot 4\pi R^2 \rho \end{cases} .$$

From this it follows that

$$\frac{\Delta M}{\Delta t} = \frac{\pi c R_g}{A},$$

and

$$\rho = \frac{1}{2AR} .$$

This remarkable expression for the flow density depends only on the distance to the mass-point singularity, with $A$ as a parameter. Its generality can be verified by integration, which yields

$$M_P = \frac{1}{A}\int_0^{R_P} 4\pi r^2 \rho dr \Rightarrow \frac{\pi R_P^2}{A} .$$

This is the same expression as the one earlier found for the relation between surface and mass for a particle, which was shown to be relevant in the microcosm and macrocosm alike.

## Expanding or non-expanding Universe?

It is possible from the model described above that the Universe may alternatively expand because it absorbs energy out of the surrounding vacuum, or maintain an equilibrium size, its expansion offset by a re-absorption of its own vacuum energy.

The latter alternative would have to include mechanisms by which the energy absorbed by the particles can be re-emitted, or radiated back into the vacuum space. Such mechanisms could, for example, be the radiation from the particles of electromagnetic energy due to oscillating electrical charges or radiation from radioactive decay, or Hawking radiation, whereby spontaneous creation of particles and anti-particles in the vacuum near black holes would lead to a radiation of energy away from the holes.

In the latter case, there must also be a mechanism by which the radiated energy can be reabsorbed by the Universe in the vacuum. In this case Hubble's constant would also come into play, this time as a combined decay and absorption constant, while the Hubble time would be the time for a particle to re-cycle its energy in the vacuum, indicated (among other things) by the observed cosmic red-shift for the photons. This is also in agreement with the mechanism described earlier, because according to the model, the total incoming vacuum flow towards each particle from the event horizon of the Universe has the same mass/energy as the particle itself. In this latter scenario the Universe may be what it is, simply because it has reached such an equilibrium situation. It could then also conform to the model for gravitation developed by Professor A. Ghosh (Ref. 3).

## References / Further reading

In general, for the well-established and known concepts of physics, reference is made to the literature and the University courses, which can be quickly found on the WEB, simply by searching the key words, such as Special Relativity, General Relativity, Schwarzschild Metric, Gravitational Redshift, Black Hole, Event Horizon, Expanding Universe, Standard Model, Planck Length, Hawking Radiation, Schroedinger Equation, Compton Wavelength, Bell Inequalities, Lorentz Transformations, Hubble Time, *etc*.

In particular, the following documents may add to the presentation by providing a complementary approach (1), adding observations (2) or theory of specific of interest (3):

1. H. Broberg. "The interface between Matter and Time - a key to understanding gravitation," in *Open Questions in Relativistic Physics*, Edited by Franco Selleri. Montreal, Apeiron, 1998. http://redshift.vif.com

2. H. Arp; *Seeing Red*. Montreal, Apeiron, 1998.
3. A. Ghosh; *Origin of Inertia*. Montreal, Apeiron, 2000

# An Alternative Picture of the Structure of Galaxies

M. Biesiada[3]
K. Rudnicki[4]
J. Syska[5]

We present some puzzling evidence about galaxies, such as their apparent motionless localization with respect to the cosmological rest frame, quantization of redshifts and the "fingers of God" phenomenon. We propose that these facts can be understood within the classical self-consistent multidimensional field theory with a dilaton. In particular, some peculiar features of emission lines from active galactic nuclei can be explained in a natural way by assuming that dilatonic configurations are placed in central parts of AGNs. Moreover, dilatonic balls in the centres of galaxies may be responsible for recently discovered kinematical effects usually attributable to hypothetical massive central black hole. We suspect that the large-scale structure in the universe could be built upon dilatonic condensates. The presence of gauge fields around dilatonic configurations could make the resulting pattern static. Properties of intrinsic redshift in dilatonic theory (steepness of $z(r)$ near centres of condensation and its domination over Doppler shifts) also make this model attractive from the point of view of the discordant redshift problem.

*Keywords*: Gravitation; Elementary particles; Active Galactic Nuclei Cosmology: theory and dark matter

## 1. Introduction

After a period of rapid growth and spectacular progress modern physics, built on the pillars of electrodynamics, General Relativity, quantum mechanics and the GSW standard model, has in many aspects approached the limits of its explanatory power. This is manifested in persisting problems with infinities in quantum field theory due to the assumption of the point-like nature of particles, problems in reconciling general relativity

[3] Department of Astrophysics and Cosmology, University of Silesia,Uniwersytecka 4, 40-007 Katowice, Poland. biesiada@camk.edu.pl; mb@imp.sosnowiec.pl
[4] Jagellonian University, Astronomical Observatory, Orla 171, Cracow, Poland
[5] Department of Astrophysics and Cosmology, University of Silesia,Uniwersytecka 4, 40-007 Katowice, Poland

with quantum mechanics (*e.g.* Einstein-Podolsky-Rosen paradox) and with understanding quantum measurement (wave-function collapse)—to mention just a few outstanding problems. During the last decade, great hopes were placed in the approach offered by the superstring theories. An excellent exposition of the potential promised by superstring theories is given by Nanopoulos (1994).

The most remarkable idea that appeared in this context is the multidimensionality of the world (the concept that our world can have more than 4 dimensions). Most of the work in this direction remained in the domain of pure thought, with the disadvantage that theoretical predictions could not be compared with observations.

It has been argued that thinking of superstring type theories in the context of elementary particles raises the problem of the inaccessibly high energies necessary to test their original predictions. On the other hand, there have been attempts to contemplate the cosmological implications of multidimensional theories. However, almost all of them were directed toward a standard inflationary, *i.e.*, big-bang, scenario (trying to provide a pre big-bang scenario), which again placed them in an untestable (*per definitionem*) domain.

If our world has more than four dimensions, the extra dimensions must not be on an equal footing with the familiar four. This problem is known as the compactification of extra-dimensions. It is still an open question as to how such compactification might take place, and what are the underlying mechanisms. Apart from simplistic visions that have appeared in the literature, one can imagine the possibility of inhomogeneous compactification, where compactified (negligibly small internal space) and noncompactified regions coexist. Recently Biesiada, Mańka & Syska (2000) found a static spherically symmetric solution of the Einstein equations in six-dimensional dilatonic theory. This solution can be thought of as a ground-state solution in the broad class of superstring-inspired theories, and also as an example of a "noncompactified domain" within the inhomogeneous compactification scenario.

The aim of this paper is twofold:

i) to recall some not widely known facts concerning the properties (*e.g.* redshifts) and distribution of galaxies in the universe, which are difficult to understand from the standard point of view.

ii) to show that these facts can be understood in the light of dilatonic theory.

Section 2 presents the above-mentioned facts from extragalactic astronomy. Then in Section 3 we digress upon the superstring inspired theories, briefly describe recently found dilatonic solution and point out applications in explaining some features of active galactic nuclei. Then, in Section 4 we broaden our vision and propose that the large-scale structure of the universe could be built upon dilatonic condensations which form a lattice-like pattern in an inhomogeneously compactified world. Finally, Section 5 presents some conclusions.

## 2. Puzzling facts about galaxies

The first puzzling fact was known as early as the first half of this century. The linear velocity of the local centroid of stars can be established in relation to the centre of our Galaxy from the Strömberg diagram, or in relation to neighbouring galaxies, from their redshifts (compare Zonn & Rudnicki 1965). Against all expectations, both velocities were found to be exactly parallel to one another (within the value of mean error) and the upper limit of the difference of their absolute values was established as 100 km/sec, implying a negligible difference in accuracy, which at that time was ~ 30 km/sec. This meant that our Galaxy is almost motionless in respect to the cosmological local rest frame, allowing a small motion parallel to the local mean transversal motion of stars in the Solar neighbourhood (roughly the orbital velocity of the Sun), within the accuracy of calculation. More accurate data now available, and new methods of establishing these velocities used by Guthrie & Napier (1996, 1993) currently give much better agreement. The upper limit for the velocity of our Galaxy with respect to neighbouring galaxies is of the order of 1 km/sec. This proves that our Galaxy is motionless, or at least almost motionless, in space.

Some newer results published in recent years seem to show that the above-mentioned characteristic feature of our Galaxy is common to many others galaxies, too. A paper by Sulentic (1993) compares the dispersion of redshifts in quintets and quartets of galaxies and the distribution of their members on the celestial sphere. The puzzling fact was found that interpreting redshifts in terms of Doppler effect as velocities of member galaxies leads to the conclusion of fast dynamical evolution of these groups, whereas the distribution of member galaxies on the celestial sphere shows no evolutionary effects. Instead, nearby and most distant quartets and quintets have approximately the same constant density of their members in projection on the celestial sphere. This points to the conclusion that tangential velocities of member galaxies in these groups are insignificantly small. The secondary conclusion is that their redshifts are probably not caused by Doppler effect.

The above-mentioned papers by Napier & Guthrie (1996, 1993) show that the values of redshifts (after elimination of the Doppler effect caused by the orbital motion of the Sun) reveal strictly discrete values within the accuracy of observations. Whatever the cause of this effect, first discovered by Tifft (1976) and called periodization or quantization of redshifts, the most important fact for our purpose is that these discrete values are not smeared in any measurable degree by a Doppler effect, which again supports the conclusion that velocities of galaxies are insignificantly small.

Also, the well known "fingers of God" effect (*e.g.*, Arp 1993) is sometimes brought up as an argument in support of the fact that redshifts of galaxies are not connected with a Doppler effect, and hence the velocities of galaxies or other extragalactic objects are in any case much lower than previously imagined. This effect has been known for several decades as the problem of missing mass in clusters of galaxies. Large dispersions of redshifts interpreted in terms of velocity dispersions would demand a larger amount of matter than actually observed.

It was, however, not easy to reconcile the above phenomena because galaxies—according to generally accepted concepts—should have huge mass concentrations in their cores subject to gravity, and thus be governed by the simple principles of mechanics. Gravitational interaction between the galaxies should cause acceleration of entire galaxies and, in effect, produce velocities of order of at least several tens of km/sec. Thus, many astronomers did not accept these observational facts, which seemed to lead to irrational conclusions.

The dilatonic hypothesis of galactic cores developed in the present paper seems to reconcile the generally accepted laws of physics with these puzzling phenomena on the assumption that the velocities of galaxies are equal to zero, or are at least insignificantly small. The possibility of explaining discrete extragalactic redshifts with dilatonic balls will be shown elsewhere.

# 3. Superstring-inspired astrophysics

Many recent ideas in theoretical physics take seriously the possibility that our world may have more than four dimensions (Green *et al.* 1987, Kaku 1988). It cannot be excluded that these theories might provide a better description of observational aspects of the world.

Until now, there has been no well-understood experimental evidence of the multidimensionality of the world, and our understanding of potential manifestations of higher dimensions is still too poor. However, attempts to seek the effects of extra dimensions in the astrophysical context have been made in the literature by Wesson (1992), Kalligas *et al.* (1995), Lim *et al.* (1992). This line of thinking is worth developing in order to gain a better understanding of possible manifestations of multidimensionality of the world. In particular, it may turn out that the effects of extra dimensions are all around us, contrary to standard expectations that inaccessibly high energies are necessary to probe the higher dimensions.

## 3.1. Dilatonic balls in six-dimensional theory

Recently a new static, spherically symmetric solution of the Einstein equations in six-dimensional dilatonic theory has been found (Biesiada, Mańka & Syska 2000). In our model we have considered a six-dimensional field theory comprising the gravitational field described by metric tensor $g_{MN}$ and a dilatonic massless scalar field $\varphi$. The field equations are obtained by extremalizing the action $\mathcal{S} = = \int \sqrt{-g}\mathcal{L} d^6 x$, or in other words, from the Euler equations for the total lagrangian $\mathcal{L}$ ($g$ denotes the determinant of the metric tensor $g_{MN}$). The total lagrangian of our model is the sum of two terms. The first one is the lagrangian for the gravitational field $\mathcal{L}_{gr} = \frac{1}{2\kappa_6}\mathcal{R}$, where $\kappa_6$ denotes the coupling constant of the six-dimensional theory analogous to the Newtonian gravity constant, and $\mathcal{R}$ is the curvature scalar of six-dimensional spacetime. The second term is the lagrangian for the dilaton field: $\mathcal{L}_\varphi = -\int \sqrt{-g}\,\frac{1}{2} g_{MN} \partial^M \varphi \partial^N \varphi d^6 x$. Note that the lagrangian for the dilaton field differs in sign from that for an ordinary massless scalar field (a typical feature of dilatons).

The presence of the dilaton in Kaluza-Klein theories is usually justified by the desire that the multidimensional Einstein equations should constitute a closed system. The dilaton is also a universal field appearing in string theories (Antoniadis, Bachas, Ellis & Nanopoulos 1988, 1989, 1991). A lagrangian like ours may come, *e.g.*, as an effective lagrangian in the six-dimensional target space from the compactified superstring theory (Ferrara 1990). There are several reasons why one may consider six-dimensional models to be attractive. Nishino & Sezgin (1984) and Salam & Sezgin (1984) have suggested that one may obtain a fermion spectrum in four dimensions within the framework of six-dimensional Kaluza-Klein supergravity. Their lagrangian was very similar to ours (differing by inclusion of boson fields). It is also curious that the number of chiral four-dimensional fermions obtained from six-dimensional Weyl spinors is finite. Our model can be thought of as providing the ground state of some larger multidimensional theory.

The action underlying our model leads to the Einstein equations

$$R_{MN} - \frac{1}{2} g_{MN} \mathcal{R} = \kappa_6 T_{MN} \tag{1}$$

coupled to the Klein-Gordon equation

$$\Box\varphi = 0 \tag{2}$$

where $R_{MN}$ is the six-dimensional Ricci tensor, $\mathcal{R}$ is the six-dimensional curvature scalar and $T_N^M$ is the energy-momentum tensor of a dilaton field $\varphi$

$$T_N^M = \partial_N \varphi \frac{\partial \mathcal{L}_\varphi}{\partial(\partial_M \varphi)} - \delta_N^M \mathcal{L}_\varphi , \tag{3}$$

and

$$\Box = -\frac{1}{\sqrt{-g}} \partial_M \left( \sqrt{-g} g^{MN} \partial_N \right).$$

One can make the following *ansatz* concerning the metric:

$$g_{MN} = \begin{pmatrix} g_{\alpha\omega} & 0 \\ 0 & g_{he} \end{pmatrix}, \tag{4}$$

where $\alpha,\omega = 0,1,2,3$ and $h,e = 5,6$. The four-dimensional sector is assumed to be spherically symmetric

$$g_{\alpha\omega} = \begin{pmatrix} e^{\nu(r)} & & & \\ & -e^{\mu(r)} & 0 & \\ & 0 & -r^2 & \\ & & & -r^2 \sin^2 \Theta \end{pmatrix}, \tag{5}$$

and the two-dimensional sector is taken as

$$g_{he} = \begin{pmatrix} -\varpi^2(r)\cos^2 \vartheta & 0 \\ 0 & -\varpi^2(r) \end{pmatrix}. \tag{6}$$

Six-dimensional coordinates $(x^M)$ are denoted by $(t,r,\Theta,\Phi,\vartheta,\varsigma)$ where $t \in [0,\infty)$ is the usual time coordinate, $r \in [0,\infty)$, $\Theta \in [0,\pi\,]$ and $\Phi \in [0,2\pi)$ are familiar three-dimensional spherical coordinates in the macroscopic space; $\vartheta \in [-\pi,\ +\pi)$ and $\varsigma \in [0,2\pi)$ are coordinates in the internal two-dimensional space; and $\varpi \in (0,\infty)$ is the "radius" of this two-dimensional internal space.

It has been shown in (Biesiada, Mańka & Syska 2000) that the following solution

$$g_{MN} = diag\left(\frac{r}{r+a}, -\frac{r}{r+a}, -r^2, -r^2\sin^2\Theta, -d^2\frac{r+a}{r}\cos^2\vartheta, -d^2\frac{r+a}{r}\right), \tag{7}$$

$$\varphi(r) = \pm\sqrt{\frac{1}{2\kappa_6}}\ln\left(\frac{r}{r+a}\right), \tag{8}$$

where $d$ is of order of $10^{-33}$ m, satisfies the coupled Einstein-Klein-Gordon equations (1), (3), (2) in the class of metrics (5), (6).

The internal space is Ricci flat. However, we must not neglect it, because its "radius" $\varpi$ is a function of $r$ and these two spaces, external and internal, are therefore "coupled." Only when $a = 0$ are these two spaces "decoupled," and our four-dimensional spacetime becomes Minkowski flat. When $a$ is not equal to zero our four-dimensional external spacetime is curved.

We now perform an asymptotic expansion of the $g_t$ metric component for $r >> a$, *i.e.*, far away from the configuration:

**Table 1**

| Example | *M* in m | a in pc |
|---|---|---|
| main sequence star | 1.0 | $10^{-13}$ |
| globular cluster | $10^4 - 10^6$ | $10^{-9} - 10^{-7}$ |
| galactic nucleus | $10^7$ | $10^{-6}$ |
| Galaxy | $5\times10^{11}$ | $2\times10^{-2}$ |

**Note**: Values of the a parameter for which the massless dilaton field in a six-dimensional world influences the dynamics of test particles in a similar way as the existence of mass *M*. All items except the third one have purely illustrative meaning. The case of "galactic nucleus" shows the value of the *a* parameter necessary to explain the central engine of galactic nuclei by dilatonic balls instead of super massive black holes.

$$g_{tt} = \frac{r}{r+a} \approx 1 - \frac{a}{r}.$$

It has a form similar to the metric induced by a mass *M* in the Newtonian limit:

$$g_{tt} \approx 1 - \frac{2GM}{c^2 r}.$$

This means that the massless scalar field (the dilaton) labelled by the *a* parameter can have the same dynamical consequences as a mass $M = ac^2/2G$. Table 1 gives the values of the *a*-parameter that mimic some astrophysically interesting masses.

In the standard derivation of the Schwarzschild solution, the free parameter in the metric tensor is identified with the total mass of a spherically symmetric configuration by requiring that, at large distances, the metric tensor should reproduce the Newtonian potential. In our case, we decided not to identify the *a*-parameter directly with *M*. The reason is that our solution describes the case where ordinary matter is absent. The only contribution to the energy-momentum tensor comes from the massless dilaton field $\varphi$.

Fig.1 illustrates the rotation curves for test particles moving along circular orbits in our model spacetime. It may appear that the weak equivalence principle is violated, since one may have different orbitals at the same radius *r* (for different values of the first integral $k_{\vartheta\varsigma}$). It is, however unclear what is the correct physical interpretation of the microscopic momenta $k_{\vartheta\varsigma}$ for macroscopic bodies. If the internal momentum ought to be fixed (*e.g.*, $k_{\vartheta\varsigma} = 0$ is

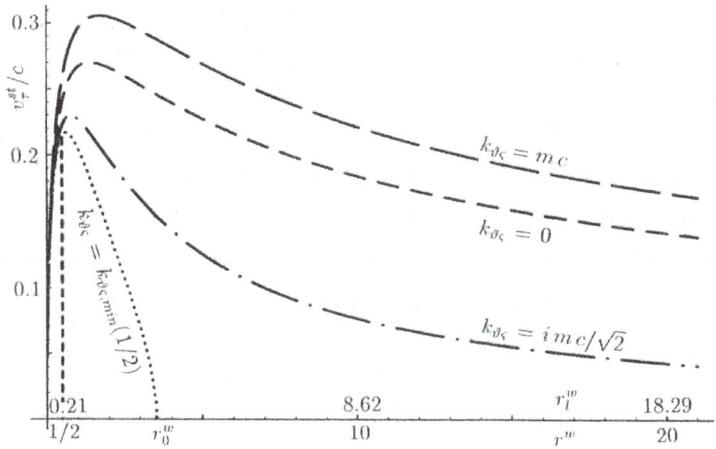


**Fig.1.** The rotation curves $v_\tau^{st}(r)$ of test particles moving in the gravitational field of spherically symmetric dilatonic configurations. The quantity $r^w$ denotes the relative radial coordinate $r^w = r/a$, *i.e.*, measured in units of the *a*-parameter. It is interesting to note that at the distances $r^w \sim 1$ relativistic orbital velocities are attained. In the peripheral regions this kinematic effect is much smaller (see the text).

distinguished since in that case, six and four-dimensional masses are equal) then there is no problem at all: the rotation curve is unique. On the other hand it is known that modern multidimensional theories could violate the equivalence principle (Cho 1992). The standard argument often invoked in this context is that the principle itself is established experimentally only within some limits (*e.g.*, derived from the accuracy of the Eötvös experiment).

Since the rotational velocity decreases with $r$ the dynamical effect of the dilaton field is maximal in regions close to the centre of spherically symmetric configurations. For example at $r \sim a$, the velocity $\upsilon_\tau$ can be as large as 0.3 $c$ (see Fig.1). Let us recall that if the dilaton field is to mimic the mass of the galactic nucleus, then $a \approx 10^{-6}$ pc, so the effect is indeed confined to very central parts in terms of galactic scales. It is interesting to note that the $a$-parameter has the same numerical value as the Schwarzschild horizon of the mass having the same dynamical effects as the dilatonic configuration in question. This means that if the dilatonic configurations (interacting with ambient matter) discussed here were realised in nature, they could easily be misinterpreted as black holes (accreting the surrounding matter).

The metric structure (7) and kinematical properties of the model are reflected in the redshift of light emitted by a test particle. The total redshift (or blueshift) can be split into two parts: a Doppler part $z_D$ associated with motion along an orbit and a gravitational part $z_{gr}$ associated with the properties of the spacetime metric (7). As stated above, one would expect these contributions to be significant in central parts of the system.

The frequency $\omega$ of light measured in units of the proper time $\tau$ ($d\tau = \sqrt{(g_{tt})}dt$) is equal to (Landau & Lifshitz 1975)

$$\omega = \frac{\omega_0}{\sqrt{g_{tt}}}, \tag{9}$$

where $\omega_0$ is the comoving frequency (in the rest frame of the source). Let us assume that a photon of frequency $\omega_s$ (measured in units of the proper time $\tau$) is emitted from the source which is located at a point $r = r_s$ where $g_{tt} = g^s_{tt}$. Then the photon is moving along a null geodesic and reaches the observer at the point $r = r_{obs}$, where $g_{tt} = g^{obs}_{tt}$, with the frequency $\omega_{obs}$

$$\frac{\omega_{obs}}{\omega_s} = \sqrt{\frac{g^s_{tt}}{g^{obs}_{tt}}}. \tag{10}$$

Let us assume for simplicity that the observer is situated at infinity. Then we get

$$\frac{\omega_{obs}}{\omega_s} = \sqrt{\frac{r^\omega_s}{r^\omega_s + 1}} \quad \text{, where } r^\omega_s = \frac{r_s}{a}, \tag{11}$$

or for the gravitational redshift

$$z_g = \frac{\lambda_{obs}}{\lambda_s} - 1 = \frac{\omega_s}{\omega_{obs}} - 1 = \sqrt{\frac{r^\omega_s + 1}{r^\omega_s}} - 1. \tag{12}$$

As might be expected, the nearer the source is to the centre of the field $\varphi(r)$, the more redshifted is the emitted photon. However, infinite redshift is attained only in the centre of the system (dilatonic configuration), and there is no singular surface like a Schwarzschild horizon.

The amplitude of Doppler redshift associated with motion along a stable orbit is equal to

$$z_D = \sqrt{\frac{c + \upsilon_r^{st}}{c - \upsilon_r^{st}}} - 1, \tag{13}$$

where the velocity $\upsilon_r^{st}$ on a stable circular orbit is

$$\upsilon_r = \frac{c^2 \mathcal{M}_\Phi}{\varepsilon_0 r(a+r)} r_l \sqrt{\frac{r+a}{r}}, \tag{14}$$

$\mathcal{M}_\Phi$ and $\varepsilon_0$ are the first integrals of motion (angular momentum and total energy), $r_l = \int_0^r dr\sqrt{-g_{rr}}$ is the physical radial distance from the centre. On Figure 2 the gravitational redshift and the amplitudes of Doppler redshifts (for trajectories with different values of $k_{\theta\zeta}$) are plotted together. One can see that up to a certain value of radial coordinate, gravitational redshifts dominate over Doppler blueshifts. In particular, the kinematical effects in the central regions $r \sim a$ where relativistically high rotational velocities ($\upsilon \sim 0.3\ c$) are attained, would be globally redshifted.

There exist observational phenomena which pose some difficulties if one tries to understand them in a simple manner within the standard concepts. For example Tanaka *et al.* (1995) have reported the detection of relativistic effects in an X-ray emission line (the $K\alpha$ line) from ionized iron in the galaxy MCG-6-30-15. The line is extremely broad, corresponding to a velocity of $\sim 10^5$ km/s $\approx 0.3\ c$, and asymmetric, with most of the line flux being redshifted. This observation is not isolated, since broad redshifted lines have been detected in several objects, but no strong blue-shifted lines have been seen (Tanaka *et al.* 1995, Mushotzky *et al.* 1995). This is an argument against any asymmetrical-outflow hypothesis in which the flow is directed away from us, because some objects should then have the flow directed toward us. On the other hand these observations can be accommodated in our model because, as shown in Fig. 2, the Doppler effect could be hidden behind a gravitational redshift, especially in central regions where relativistic $\upsilon_\tau \sim 0.1\ c$ orbital velocities are attained. If one tries to reconcile the profiles of the X-ray iron emission lines with the standard black hole scenario, one has to conclude that much of the emission originates from within $6r_g$ (Iwasawa *et al.* 1996a, 1996b). This means, in particular, that in a standard scenario the black hole ought to be spinning rapidly in order to allow the disk to extend into such central regions by frame dragging. Our model provides an

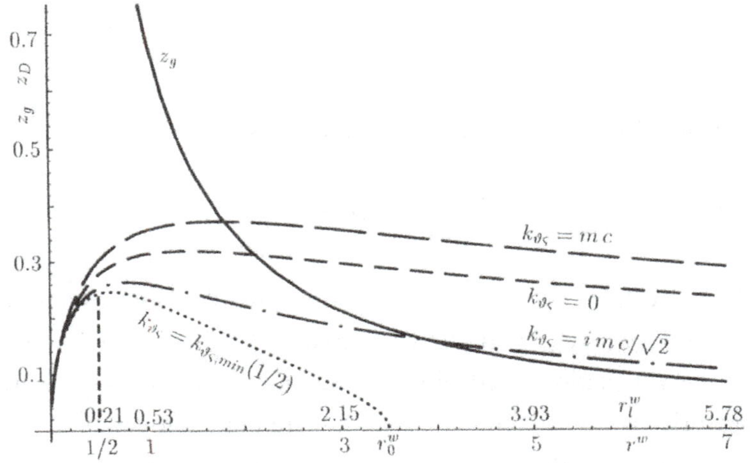


**Fig.2.** The amplitudes of Doppler shift (red or blue-shift) $z_D$ caused by motion of a particle along a stable circular orbit for different values of internal momentum. The curve $z_g$ denotes the gravitational redshift of the dilatonic configuration discussed here. It can be seen that the Doppler effect is always hidden behind the gravitational redshift up to a certain value of the radial coordinate.

alternative scenario for explaining the gross features of redshifted iron emission lines in Seyfert galaxies.

Besides the problems with active galactic nuclei, there is strong kinematical evidence that at least some galaxies have dark objects causing strong gravitation in their centres (Kormendy & Richstone 1995). The common practice is to identify them with black holes, although some authors honestly admit that all the stellar-dynamical interactions invoked in favour of their existence only require a dark object causing gravitation (Tremaine 1997). As the HST mission supplies more accurate data, the evidence is mounting for the existence of central dark objects causing gravitation in galactic nuclei. Our dilatonic balls, which cause attraction similar to gravitation or, simply put, cause gravitation without mass, can provide an alternative explanation for the massive central dark objects. This idea can be placed in a broader perspective, as envisaged in the next section.

## 3.2 Broadening vision: dark matter and large scale structure

In light of the discussion from section 2, the crucial question is whether it is possible that dilatonic configurations that have seeded galaxies can be maintained static with respect to one another. There appears to be hope that the above question can be answered in the affirmative.

The action leading to our dilatonic solution is motivated (in the sense of being typical) by superstring theories. A more general action in the above-mentioned class of theories leads to the following four-dimensional effective lagrangian:

$$\mathcal{L}=\sqrt{-g}\begin{bmatrix}\frac{1}{2\kappa_6}\mathcal{R}-\frac{1}{2}g^{\mu\nu}\partial_\mu\varphi\partial_\nu\varphi \\ -\frac{1}{2}\exp(-2\lambda\varphi)g^{\mu\nu}\partial_\mu\alpha\partial_\nu\alpha-\frac{1}{4}\exp(\lambda\varphi)F^j_{\mu\nu}F^{\mu\nu}_j \\ +\frac{1}{64\pi^2 f_{PQ}}\exp(-\lambda\varphi)\in^{\mu\nu\rho\sigma}\alpha F^i_{\mu\nu}F_{\rho\sigma i}\end{bmatrix}, \quad (15)$$

where the second scalar field $\alpha$ is an axion field, $F^a{}_{\mu\nu}$ are the field strengths of the gauge fields $A^a{}_\mu$ (for details see Dick 1997).

Hence our dilatonic solution can be thought of as a ground state for the more general case when the dilatonic configurations are "dressed" with other fields, which can be another scalar (axionic) as well as charged gauge fields.

The gravitational attraction due to gravity $g_{\mu\nu}$ and the massless scalar fields $\varphi$ and $a$ can be exactly cancelled by a repulsion due to charged gauge fields (Duff *et al.* 1999). The resulting static configuration is analogous to the famous static Majumdar-Papapetrou solution (Majumdar 1947, Papapetrou 1947) describing two (oppositely) charged gravitationally interacting black holes in classical General Relativity. A particular lattice solution of this kind has recently been discussed by Duff *et al.* (1999) in the context of the *M*-theoretical explanation of reported striking regularities of the large-scale structure in the universe.

Since the 70's there has been no doubt that the motions of outlier stars and clouds in galaxies are not compatible with the observed luminous matter distribution. It is also rather well established that one cannot solve the dark matter puzzle by ordinary baryonic matter alone (see *e.g.*, Kolb & Turner 1989). It seems that the solution may lie in the so-called cold dark matter represented by some kind of weakly interacting massive particles (see *e.g.* Sikivie 1994).

Our model may be thought of as a starting point for more detailed investigations of the non-baryonic dark matter scenario. As we see here, the dilaton (massless scalar field) may act dynamically in the same way as massive bodies. In this way the dilaton can become another candidate for the cold dark matter. On the galactic scale, however, where dilatonic configurations could be invoked to explain the central engine for AGNs and quasars, our model does not solve the flat rotation curve problem. In this aspect it would be very interesting to extend the model discussed in this paper by inclusion of an axion field. Axions are predicted from the Peccei-Quinn mechanism to solve the strong CP problem in QCD (Peccei & Quinn 1977), but they also appear in string theory (Green *et al.* 1987). On the other hand, string theory also predicts a dilaton (similar to Kaluza-Klein theories): the evidence is mounting that the axion and dilaton should come together (Dick 1997). The behaviour of an axion field in a background dilatonic model presented above will be investigated elsewhere.

## 4. Conclusions and perspectives

We may imagine that our six-dimensional world could be compactified in an inhomogeneous manner. In this picture dilatonic configurations would form a kind of a condensate. It is interesting to notice that there has recently been a tendency to draw analogies between cosmology and condensed matter physics (Hu 1988, 1996, Smolin 1995). This idea would be particularly fruitful if it turned out that general relativity were an emergent macroscopic theory originating as a low energy collective state of some more fundamental theory. The dilatonic "centres of condensation" could be the seeds for the large-scale structure of the Universe (galaxies). If we look at the dilaton as a candidate for dark matter, the above-mentioned condensations can be regarded as a specific version of the cold dark-matter scenario, which is known to be a successful scenario for the emergence of the large-scale structure in the Universe. Although it is very speculative at this stage, the localized character of dilatonic configuration has an advantage over the other CDM scenarios, where the elusive tiny particles (such as neutralinos, axions, *etc.*) have to contribute by producing clumps heavy enough to seed the galaxies. On the other hand there is evidence (Bahcall *et al.* 1995) that most dark matter in the universe resides in large dark halos around galaxies. This means that the total masses of clusters and superclusters can be accounted for by the total mass (*i.e.*, including dark halos) of their members plus intergalactic hot gas (for details see Bahcall *et al.* 1995). Extensive galactic modelling (Gates *et al.* 1995) constrained by the data from the MACHO search by gravitational lensing (Alcock *et al.* 1995) indicates that, in most viable models, only about 30% of dark halo mass could be attributable to compact baryonic dark matter. Hence one should seriously consider the possibility that a significant fraction of the galactic dark matter might be nonbaryonic. One of the primary candidates in this class is an axion. However, stringent constraints imposed on the mass of an axion by stellar evolution and the SN 1987 neutrino burst (Kolb & Turner 1989) raise the question of how to keep such light and weakly interacting (with ordinary matter) particles bound to the galactic halo.

As mentioned earlier, we believe that the coupled system of an axion and a dilaton (axion propagating in the dilatonic background) is promising in this respect. That they should come together is predicted by particle physics. In our picture, an essentially classical dilatonic configuration would provide a background to anchor axions in the form of an extended halo. Details of this model will be presented in a separate paper.

There exists spectroscopic evidence that galaxies seem not to move with respect to each other under mutual gravity. We have seen that in the framework of string-inspired theories one can imagine a static, structured universe. The centres of condensation, which act dynamically as massive bodies, are static with respect to each other, which could explain all the puzzling phenomena mentioned in Section 2, but baryonic mass (stars, gas and dust) captured by them behave in a normal way, *i.e.*, are moving in the potential wells provided by the condensations.

Macroscopic dilatonic configurations (which are transparent to light) can produce large redshifts in their central parts. We have suggested that this phenomenon can explain certain properties of emission lines from active galactic nuclei. However, it can also point toward a correct understanding of puzzling properties of the redshift which, although neglected by the mainstream, have been reported repeatedly (Arp 1993 and references therein). Let us imagine that dilatonic condensates are produced with some spectrum of the $a$-parameter. If it can happen that at some place $a \approx 100$ kpc, then because of the steepness of the $z_g(r)$ function, the redshift can be seen to change from $z_g = 3$ to $z_g = 0.4$ at a distance of about 50 kpc. This can resolve a long-standing puzzle of discordant redshifts in quasar-galaxy associations connected by material bridges. What superstring theory says about the possible spectrum of $a$ is thus a challenging question. It is indeed remarkable that ideas which emerged from a desire to understand elementary particles may turn out to explain the fundamental features of the universe in the large scale.

## References

Alcock, C., *et al.*., 1995, *Astrophys. J.* 445, 133

Antoniadis I., Bachas C., Ellis J., Nanopoulos D.V., 1988, *Phys.Lett.B* 211, 393

Antoniadis I., Bachas C., Ellis J., Nanopoulos D.V., 1989, *Nucl. Phys.B* 328, 117

Antoniadis I., Bachas C., Ellis J., Nanopoulos D.V., 1991, *Phys.Lett.B* 257, 278

Arp H., 1993, Fitting Theory to Observation—From Stars to Cosmology. In: Arp H.C., Keys C.R., Rudnicki K. (eds.) *Progress in New Cosmologies*, Plenum Press, New York, p.1

Bahcall N., Lubin L.M., Dorman V., 1995, *ApJ*, 447, L81

Biesiada M., Mańka R., Syska J., 2000 "A new static spherically symmetric solution in six-dimensional dilatonic Kaluza-Klein theory" *Int.J.Mod.Phys.D* **9**, 71-78 (2000).

Cho Y.M., 1992, *Phys.Rev.Lett.* 68, 3133

Dick, R., 1997, The dilaton as a candidate for dark matter. hep-th/9609190 In: *Dark Matter in Astro and Particle Physics*, eds. H.V. Klapador-Kleingrothaus and Y.Ramachers, World Scientific: Singapore, pp. 395--402

Duff M.J., Hoxha P., Lü H., Martinez-Acosta R.R., Pope C.N., 1999, A Lattice Universe from M-theory. astro-th/9712301 *Phys.Lett.B* 451, 38-45

Ferrara S., in *1990 Summer School in High Energy Physics and Cosmology*, eds. J.C.Pati, S.Randjbar-Daemi, E.Sezgin, Q.Shafi, World Scientific, Singapore, pg.111-140, 1991

Gates E., Gyuk G., Turner M., 1995, *Phys.Rev.Lett.* 74, 3724

Green M.B., Schwarz J.H., Witten E., 1987, *Superstring Theory*. Cambridge University Press, Cambridge

Guthrie B.N.G., Napier W.M. 1993, Redshift Periodicity in the Local Supercluster. In: Arp H.C., Keys C.R., Rudnicki K. (eds.) *Progress in New Cosmologies*, Plenum Press, New York, p.29

Guthrie B.N.G., Napier W.M. 1996 *A&A* 310, 353 .

Hu B.L., 1988, Cosmology as a 'Condensed Matter' Physics. In: Chan Y.W. *et al.* (eds.) *Proc. 3rd Asia Pacific Physics Conference.* World Scientific, Singapore, p.301.

Hu B.L., 1996, *General Relativity as Geometro-Hydrodynamics.* gr-qc/9607070

Iwasawa K., Fabian A.C., Mushotzky R.F., Brandt W.N., Awaki H., Kunieda H., 1996, *MNRAS* 279, 837

Iwasawa K. *et al.*, 1996, *MNRAS* 282, 1038

Kaku M., 1988, *Introduction to Superstrings.* Springer, New York

Kalligas D., Wesson P., Everitt C.W.F., 1995, *ApJ* 439, 548--557

Kolb E., Turner M., 1989, *The Early Universe.* Addison-Wesley, New York

Kormendy J., Richstone D.O., 1995, *ARA&A*, 33, 581

Landau L.D., Lifshitz E. M., 1975, *The Classical Theory of Fields.* Pergamon, New York

Lim P., Wesson P., 1992, *ApJ*, 397, L91

Majumdar S.D., 1947, *Phys.Rev.* 72, 390

Mushotzky R.F. *et al.*., 1995, *MNRAS* 272, L9-L12

Nanopoulos D.V., 1994, *Riv. Nouvo Cim.* 17, 1

Nishino M., Sezgin E., 1984, *Phys.Lett.B* 144, 187

Salam A., Sezgin E., 1984, *Phys.Lett.B* 147, 47

Papapetrou A., 1947, *Proc. R. Irish Acad.* A51, 191

Peccei R.D., Quinn, H.R., 1977, *Phys.Rev.Lett.* 38, 1440; 1977, *Phys.Rev.D* 16, 1791

Sikivie, P., "Sources and Distributions of Dark Matter," in eds. L.Bergstrom, P.Carlson, P.O.Hulth and H.Snellman (1994) Proc. of the Conference *Trends in Astroparticle Physics* Stockholm

Smolin L., 1995, *Cosmology as a Problem in Critical Phenomena.* gr-qc/9505022

Sulentic J., 1993 Compact Groups of Galaxies. In: Arp H.C., Keys C.R., Rudnicki K. (eds.) *Progress in New Cosmologies*, Plenum Press, New York, p.49

Tanaka Y. *et al.*, 1995, *Nature* 375, 659

Tifft W.G., 1976, *Astrophys.J* 206, 38.

Tremaine S., 1997, In: Bahcall J.N., Ostriker J.P.(eds.) *Unsolved Problems in Astrophysics.* Princeton University Press, Princeton.

Wesson P., 1992, *ApJ* 394, 19

Zonn W., Rudnicki K., 1965, *Stellar Astronomy.* Clearing House for Federal Scientific and Technical Information, Washington D.C.

# Electromagnetism and Cosmology

Edward Kapuścik[6]

A new form of the generally covariant Maxwell electrodynamics is considered. The theory may be applied to arbitrary media without explicit knowledge of the constitutive relations of the media. This opens new possibilities for the description of electromagnetism in the Universe as a whole.

## Introduction

It is generally believed that Maxwell electrodynamics is the best theory of electromagnetism.[1] It is therefore a good candidate for a theoretical description of all electromagnetic phenomena in the whole Universe. There are, however, at least two conditions to be fulfilled for that purpose. First, the Maxwell electrodynamics must be formulated in a generally covariant way, because all phenomena in the Universe must be described covariantly. Second, the theory must be applicable to arbitrary media, since at present we do not know the exact shapes of the constitutive relations for all objects in the Universe. Unfortunately, at present the general covariant form of Maxwell electrodynamics is mostly known only for the vacuum-like media in which the constitutive relations of the vacuum type are satisfied.[2] For a general medium we need to know a fourth-order antisymmetric tensor density to describe the medium in the linear approximation. For non-linear media the theory is not developed.

The aim of this note is to show that there exists a special reformulation of Maxwell theory which satisfies both of the above criteria.

## Electromagnetism and general relativity

The unification of electromagnetism with gravity was not just the dream of Albert Einstein. One of the reasons why this problem could not be solved in

[6] Department of Theoretical Physics, Henryk Niewodniczański Institute of Nuclear Physics, ul. Radzikowskiego 152, Kraków 31 342, and Department of Theoretical Physics, University of Lódź, ul. Pomorska 149/153, Lódź 90 236, Poland

the past is the notorious lack of generally covariant models of Maxwell electrodynamics in arbitrary media.

Recently we have shown[3,4] that the formulation of Maxwell electrodynamics in terms of the fields $\vec{D}(\vec{x},t)$ $\vec{H}(\vec{x},t)$ $\vec{P}(\vec{x},t)$ and $\vec{M}(\vec{x},t)$, instead of the fields $\vec{E}(\vec{x},t)$, $\vec{B}(\vec{x},t)$, $\vec{D}(\vec{x},t)$ and $\vec{H}(\vec{x},t)$, makes it possible to avoid all the problems connected with the covariant form of the constitutive relations. The role of the constitutive relations is partially taken by a new completely antisymmetric tensor field of the third rank. In this reformulation of Maxwell electrodynamics, the whole content of electrodynamics is contained not in one set of Maxwell equations, but in two such sets. The difference between the standard formulation and the new one consists in the fact that one set of field equations contains only the fields $\vec{D}(\vec{x},t)$ and $\vec{H}(\vec{x},t)$, while the second set of field equations contains the fields $\vec{P}(\vec{x},t)$ and $\vec{M}(\vec{x},t)$. All field equations are inhomogeneous with two conserved vector currents and one tensor source field which is used to formulate the Faraday induction law in an arbitrary medium. More precisely, the two sets of Maxwell equations are of the form

$$rot\,\vec{D}(\vec{x},t)=\frac{1}{c^2}\frac{\partial\vec{H}(\vec{x},t)}{\partial t}-\frac{1}{c^2}\vec{j}_M(\vec{x},t) \tag{1}$$

$$div\,\vec{H}(\vec{x},t)=\rho_M(\vec{x},t) \tag{2}$$

$$rot\,\vec{H}(\vec{x},t)=\frac{\partial\vec{D}(\vec{x},t)}{\partial t}+\vec{j}_M(\vec{x},t) \tag{3}$$

$$div\,\vec{D}(\vec{x},t)=\rho_M(\vec{x},t) \tag{4}$$

and

$$rot\,\vec{P}(\vec{x},t)=\frac{1}{c^2}\frac{\partial\vec{M}(\vec{x},t)}{\partial t}-\frac{1}{c^2}\vec{j}_M(\vec{x},t) \tag{5}$$

$$div\,\vec{M}(\vec{x},t)=-\rho_M(\vec{x},t) \tag{6}$$

$$rot\,\vec{M}(\vec{x},t)=-\frac{\partial\vec{P}(\vec{x},t)}{\partial t}+\vec{j}_P(\vec{x},t) \tag{7}$$

$$div\,\vec{P}(\vec{x},t) = -\rho_P(\vec{x},t) \tag{8}$$

where $\rho(\vec{x},t)$ and $\vec{j}(\vec{x},t)$ are the usual densities of external charge and current, respectively, while $\rho_P(\vec{x},t)$ and $\vec{j}_P(\vec{x},t)$ are the usual densities of polarized charge and current, respectively. The nature of the new scalar and vector densities $\rho_M(\vec{x},t)$ and $\vec{j}_M(\vec{x},t)$ has been clarified in reference 4.

The electromagnetic fields $\vec{D}(\vec{x},t)$, $\vec{H}(\vec{x},t)$, $\vec{P}(\vec{x},t)$ and $\vec{M}(\vec{x},t)$ have the same meaning as in the standard Maxwell equations. For distribution-valued sources, as in the case of point charges, all fields are generalized functions. When the new set of Maxwell equations is solved, they may be used to define the standard vacuum electromagnetic fields $\vec{E}(\vec{x},t)$ and $\vec{B}(\vec{x},t)$ with the usual formulae

$$\vec{E}(\vec{x},t) = \frac{1}{\varepsilon_0}\left[\vec{D}(\vec{x},t) - \vec{P}(\vec{x},t)\right] \tag{9}$$

and

$$\vec{B}(\vec{x},t) = \mu_0\left[\vec{H}(\vec{x},t) + \vec{M}(\vec{x},t)\right], \tag{10}$$

where $\varepsilon_0$ and $\mu_0$ are the standard electromagnetic constants of the vacuum. For distribution-valued sources these fields must have the mathematical properties of smooth test functions in the sense of generalized function theory.[3]

As is well known, the generally covariant form of macroscopic electrodynamics has been found only for the vacuum case.[2] In this formulation two objects are utilized: the antisymmetric tensor field $F_{\mu\nu}(x)$ and the antisymmetric tensor density $\mathcal{H}^{\mu\nu}(x)$, which satisfy the generally covariant Maxwell equations

$$\partial_\mu F_{\nu\lambda} + \partial_\nu F_{\lambda\mu} + \partial_\lambda F_{\mu\nu} = 0, \tag{11}$$

$$\partial_\mu \mathcal{H}^{\mu\nu} = -\mathcal{J}^\nu, \tag{12}$$

where $\mathcal{J}^\nu$ is the vector density of external charges and currents. In the linear approximation the basic electromagnetic fields may be related by the generally covariant constitutive relation

$$\mathcal{H}^{\mu\nu}(x) = \eta^{\mu\nu\lambda\rho}(x) F_{\lambda\rho}(x) \tag{13}$$

where the tensor density $\eta^{\mu\nu\lambda\rho}(x)$ describes the electromagnetic properties of the medium. Due to several symmetry properties, the tensor density $\eta^{\mu\nu\lambda\rho}$ contains only 20 independent components. These components, in the case of the vacuum, are usually expressed by the metric tensor $g_{\mu\nu}(x)$. As a result, the relation (13) takes the form

$$\mathcal{H}^{\mu\nu}(x) = \mu_0 \sqrt{-g}\, g^{\mu\lambda}(x) g^{\nu\rho}(x) F_{\lambda\rho}(x), \tag{14}$$

or conversely

$$F_{\mu\nu}(x) = \frac{1}{\mu_0\sqrt{-g}} g_{\mu\lambda} g_{\nu\rho} \mathcal{H}^{\lambda\rho}(x), \tag{15}$$

where as usual,

$$g = \det(g_{\mu\nu}). \tag{16}$$

In the new reformulation of electrodynamics the left hand sides of all Maxwell equations in the sets (1)-(8) *for arbitrary media* have exactly the same form as in the standard vacuum Maxwell equations. From this it follows that the only possible version of their generally covariant form is

$$\partial_\mu F^{(D,H)}_{\nu\lambda} + \partial_\nu F^{(D,H)}_{\lambda\mu} + \partial_\lambda F^{(D,H)}_{\mu\nu} = j_{\mu\nu\lambda} \tag{17}$$

$$\partial_\mu \mathcal{H}^{\mu\nu}_{(D,H)} = \mathcal{J}^\nu \tag{18}$$

$$\partial_\mu F^{(P,M)}_{\nu\lambda} + \partial_\nu F^{(P,H)}_{\lambda\mu} + \partial_\lambda F^{(P,M)}_{\mu\nu} = j_{\mu\nu\lambda} \tag{19}$$

$$\partial_\mu \mathcal{H}^{\mu\nu}_{(P,M)} = \mathcal{J}^\nu_P \tag{20}$$

where the corresponding tensor fields and tensor densities are constructed form the pairs of fields $(D,H)$ and $(P,M)$ according to the content of the parentheses. Clearly, each tensor field is related to its corresponding tensor density by the same relations as in (14) and (15).

For linear media the polarized charge and current densities are linearly related to the external charges and currents. In our case, this means that all the polarization and magnetization properties of such media are described by the relation

$$\mathcal{J}^{\mu}_{P}(x)=\varepsilon^{\mu}_{\nu}(x)\mathcal{J}^{\nu}(x), \tag{21}$$

where $\varepsilon^{\mu}_{\nu}(x)$ is the polarization and magnetization mixed tensor of the medium. It contains 16 independent components. The remaining 4 functions allowed by the general form of the material tensor density $\eta^{\mu\nu\lambda\rho}(x)$ are provided by the 4 independent components of the totally antisymmetric tensor $j_{\mu\nu\lambda}(x)$ formed from $\rho_M(x)$ and $\vec{j}_M(x)$. It is erroneous to try to relate this antisymmetric tensor to the external current, as in (21), because this will introduce an additional 16 new functions. It is also erroneous to treat this tensor as any kind of current, because it is not a tensor density. It is therefore impossible to obtain from it global covariant quantities by the process of integration. The only physically correct interpretation is to treat the source terms in eqs. (17) and (19) as terms describing the influence of the medium on the Faraday induction law. Such terms are absent in the case of the vacuum.

The relations between electromagnetic tensors $F^{(D,H)}_{\mu\nu}(x)$ and $F^{(P,M)}_{\mu\nu}(x)$ and tensor densities $\mathcal{H}^{\mu\nu}_{(D,H)}$ and $\mathcal{H}^{\mu\nu}_{(P,M)}$ of the type (14)-(15) create serious physical problems in the case when the metric tensor $g_{\mu\nu}(x)$ is treated as the gravitational field. The problem is in the choice of which electromagnetic field is the primary elementary field and which one is the composite field made up of electromagnetic and gravitational fields. We must remember here that in field theories, every product of fields must be treated as a composite field, and the relations (14)-(15) evidently introduce composite fields. Both choices are unsatisfactory because both basic electromagnetic fields should be as fundamental as the gravitational field.

The solution of this problem lies in the introduction of the electromagnetic potentials. Due to the fact that, in our case, all Maxwell equations are inhomogeneous we must introduce more potentials than are usually dealt with. Taking into account that only part of the electromagnetic fields are tensors and the others are tensor densities, we must introduce the customary vector potentials $A^{(D,H)}_{\mu}$ and $A^{(P,M)}_{\mu}$ and new potentials $\mathcal{A}^{\mu\nu\lambda}_{(D,H)}$ and $\mathcal{A}^{\mu\nu\lambda}_{(P,M)}$ which are antisymmetric tensor densities of third rank. Treating these quantities as primary electromagnetic fields we arrive at the following representations of the customary electromagnetic tensor and tensor density fields

$$F_{\mu\nu}=\partial_{\mu}A_{\nu}-\partial_{\nu}A_{\mu}+(-g)^{-1/2}g_{\mu\lambda}g_{\nu\rho}\partial_{\sigma}\mathcal{A}^{\sigma\lambda\rho} \tag{22}$$

$$\mathcal{H}^{\mu\nu} = \partial_\lambda \mathcal{A}^{\lambda\mu\nu} + (-g)^{1/2} g^{\mu\lambda} g^{\nu\rho} \left(\partial_\lambda A - \partial_\rho A_\lambda\right) \tag{23}$$

where we have omitted the corresponding sub- and superscripts because the relations of the electromagnetic fields to the corresponding potentials are the same in the cases of fields of the types $(D,H)$ and $(P,M)$.

The old and new potentials are not unique. They undergo the following gauge transformations

$$A_\mu \to A_\mu + \partial_\mu f \tag{24}$$

$$\mathcal{A}^{\lambda\mu\nu} \to \mathcal{A}^{\lambda\mu\nu} + \partial_\sigma a^{\sigma\lambda\mu\nu} \tag{25}$$

where $f(x)$ and $a^{\sigma\lambda\mu\nu}(x)$ are the scalar and tensor density gauging quantities. Here $a^{\sigma\lambda\mu\nu}(x)$ is a totally antisymmetric tensor density of the fourth rank.

The basic electromagnetic fields $A_\mu(x)$ and $\mathcal{A}^{\lambda\mu\nu}(x)$ satisfy the following field equations

$$\partial_\mu \left[(-g)^{1/2} g^{\mu\lambda} g^{\nu\rho} \left(\partial_\lambda A_\rho - \partial_\rho A_\lambda\right)\right] = \mathcal{J}^\nu, \tag{26}$$

$$\partial_\lambda \left[(-g)^{1/2} g_{\mu\alpha} g_{\nu\beta} \partial_\sigma \mathcal{A}^{\sigma\alpha\beta}\right] + cycl.in(\lambda,\mu,\nu) = j_{\lambda\mu\nu}. \tag{27}$$

These equations show that the basic and independent electromagnetic fields $A_\mu(x)$ and $\mathcal{A}^{\lambda\mu\nu}(x)$ propagate in spacetime only as aggregates of composite fields formed together with the gravitational field. The same situation will arise when we take any gauge conditions for the electromagnetic potential. Both field equations and gauge conditions reflect specific properties of the interaction between gravity and electromagnetism. The gauge conditions are, therefore, as physical as the field equations are.

Before closing this discussion let us observe that the vector density of current $\mathcal{J}^\mu$ is always written in the form

$$\mathcal{J}^\mu = \sqrt{g} J^\mu,$$

where $J^\mu$ is the vector field of the electromagnetic current. This means that, again, we are dealing with a composite field, and at the very beginning the interactions have been divided into pure gravitational (described by the metric tensor) and pure electromagnetic interactions (described by the vector field of current). The electromagnetic fields do not contain such a division because they are determined by the Maxwell equations, and the solutions of differential equations do not follow the separability properties of the source terms. This situation must, however, be regarded as highly unsatisfactory because it creates an obstacle to unifying electromagnetism with gravity. We must bear in mind that, in the construction of the extremely successful standard model of electroweak interactions, we have gauge fields, which create the electromagnetic sector and the weak sector only after taking their proper superpositions. This means that the correct way to unification in the case of electromagnetism and gravity should also start with quantities that are neither electromagnetic nor gravitational. All basic fields should be elementary at the beginning of the construction. The identification of the electromagnetic and gravitational sectors should come at the end, when we define the composite fields that will express physical laws of electromagnetism and gravity. This means, however, that we should start with forms of the source terms in our new sets of Maxwell electrodynamics that are different from (28). Unfortunately, a satisfactory solution to this problem has not yet been found.

As a final remark, we wish to stress the fact that, in the approach presented here, we have used all the mathematical quantities provided by the mathematics of arbitrary four-dimensional manifolds without using the notion of covariant derivative.[5] This is an advantage of our approach, because any use of the covariant derivative always introduces additional interactions of electromagnetism with gravity.

## Conclusions

We have shown that Maxwell electrodynamics in its generally covariant form may be applied to arbitrary media. This creates some hope that it may also be applied to the matter present in all parts of the Universe. It also may be used to formulate cosmological principles which will take into account the electromagnetic properties of the Universe as a whole.

## References

1. See, for example, the standard textbook by J. D. Jackson; *Classical Electrodynamics*, John Wiley and Sons Inc., N.Y., 1962
2. E. J. Post; *Formal Structure of Electromagnetics*, North-Holland Publishing Company, Amsterdam, 1962
3. E. Kapuścik; *Comm. JINR* E2-91-272, Dubna, 1991.

4. E. Kapuścik; "Generally Covariant Electrodynamics in Arbitrary Media," in: *Open Questions in Relativistic Physics*, pp. 227-231, ed. F. Selleri, Montreal, Apeiron, 1998.
5. E. Schroedinger; *Spacetime Structure*, Cambridge University Press, 1950.

# Space and Time should be Preferred to Spacetime - 1

F. Selleri[7]

Transformations of space and time between inertial systems are set up by starting from two empirically based assumptions: (1) The two-way velocity of light is the same in all inertial systems; (2) Clock retardation takes place with the usual velocity-dependent factor when clocks move with respect to an isotropic reference frame. The transformations thus obtained contain a free parameter $e_1$, the coefficient of $x$ in the transformation of time. The Theory of Special Relativity is recovered for a particular choice of $e_1$. Different values of $e_1$ correspond to different theories of space and time, which are to a large extent empirically equivalent. We show that Michelson type experiments, aberration, occultation of Jupiter's satellites, and radar ranging of planets are insensitive to the choice of $e_1$. Several other experiments lead to the same conclusion. An exception is discussed in Part II.

## 1. Introduction

The one-way velocity of light has never been measured. The obstacles lie not in practical difficulties, but in the relativistic question of clock synchronization, as discussed by Reichenbach [1], Jammer [2], Mansouri-Sexl [3] and Will [4]. Following the opinion often expressed by Poincaré [5] and Einstein [6], all one-way velocities have been considered devoid of physical interest. In fact, if the one-way velocity of one physical system were known, the velocity of any other system could be determined with a single clock by measuring the time difference between arrivals of the two systems that started simultaneously from the same point.

One-way velocities would, however, appear to be rather natural properties, after all, given that light and other objects go from one point to another in well-defined ways in nature. In an attempt to clarify this matter, recently a set of "equivalent" theories has been defined [7] differing from one another in clock synchronization only. The set includes the Theory of Special Relativity (TSR). They are briefly reviewed below. *A priori* one would hope to

[7] Università di Bari, Dipartimento di Fisica, INFN - Sezione di Bari, Via Amendola 173, I-70126 Bari, Italy

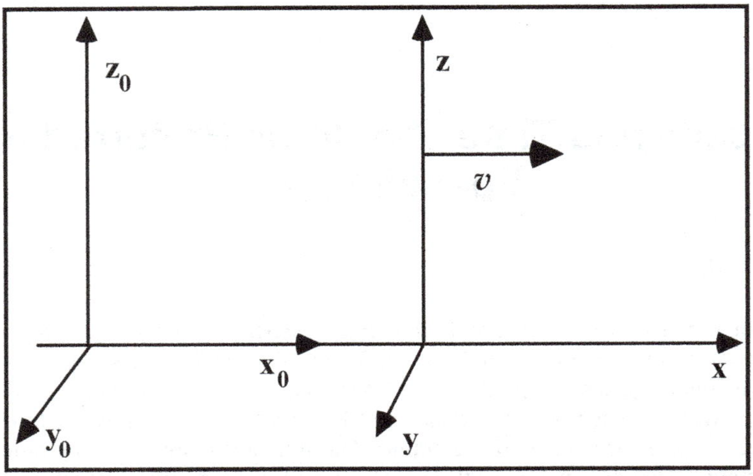


**Figure 1**. An inertial system $S$ in which orthogonal Cartesian coordinates $(x, y, z)$ have been set up is moving with velocity $v$ with respect to another inertial system $S_0$ similarly endowed with the Cartesian coordinates $(x_0, y_0, z_0)$.

find empirical differences between these theories, but the following experiments, considered fundamental for the TSR, are explained equally well by all the theories in the set: Michelson type, aberration of starlight, occultation of Jupiter's satellites, and radar ranging of planets.

In the second (review) paper presented at this conference it is shown that a measurable quantity $\rho$ exists for which TSR predicts $\rho = 1$ relative to all inertial frames. Under extremely general conditions we will prove that $\rho = (c + \upsilon)/(c - \upsilon)$ for a class of rotating frames having the same peripheral velocity $\upsilon$ and arbitrarily small acceleration $a$. Thus, TSR gives rise to a discontinuity. The limit $a \to 0$ should instead be smooth, because our empirical knowledge about inertial reference frames is obtained in laboratories with nonzero acceleration, *e.g.*, because of the Earth's rotation. Only one theory of the equivalent set gives a continuous transition between accelerated and inertial reference frames, and it is not TSR.

## 2. The General Transformations

Given the inertial frames $S_0$ and $S$ one can always choose two systems of Cartesian orthogonal coordinates (see Fig. 1) by assuming:

($i$) that space is homogeneous and isotropic, and that time is homogeneous, at least if judged by observers at rest in $S_0$.

(*ii*) that in $S_0$ the velocity of light has the same value "*c*" in all directions, so that clocks can be synchronized in $S_0$ and velocities relative to $S_0$ can be measured.

(*iii*) that the origins of $S$ and $S_0$ coincide for $t = t_0 = 0$.

(*iv*) that the origin of $S$, observed from $S_0$, is seen to move with velocity $\upsilon$ parallel to the $+x_0$ axis, that is, according to the equation $x_0 = \upsilon t_0$.

(*v*) that planes $(x_0, y_0)$ and $(x, y)$ are superimposed at all times $t_0$.

The resulting geometrical configuration is that of Fig. 1.

Concerning assumptions (*i*) and (*ii*), notice that they are unobjectionable from the standpoint either of relativity or of any plausible hypothesis involving a privileged frame; for relativity they are true for all inertial systems, and in the latter case they can be assumed for the privileged system itself.

It was shown in [7] that the above conditions reduce the transformation laws from $S_0$ to $S$ to the form

$$\begin{cases} x = f_1 (x_0 - \upsilon t_0) \\ y = g_2 y_0 \\ z = g_2 z_0 \\ t = e_1 x_0 + e_4 t_0 \end{cases}, \tag{1}$$

where the multiplying factors $f_1$, $g_2$, $e_1$, $e_4$ are expected to depend on the velocity $\upsilon$ of $S$ measured in $S_0$. The one-way velocity of light $c(\theta)$ relative to $S$, in a direction forming an angle $\theta$ (in $S$) with the $x$-axis (Fig. 1) can be obtained from (*ii*) above and from (1)[7]. Its inverse turns out to be

$$\frac{1}{c(\theta)} = \left[\frac{e_1}{f_1} + \frac{R\beta}{cf_1(1-\beta^2)}\right]\cos\theta + \frac{R}{c}\left[\frac{\cos^2\theta}{f_1^2(1-\beta^2)^2} + \frac{\sin^2\theta}{g_2^2(1-\beta^2)}\right]^{1/2}, \tag{2}$$

where $\beta = \upsilon / c$ and

$$R = e_1 \upsilon + e_4 . \tag{3}$$

Consider now the two-way velocity of light $c_2(\theta)$ from any point $A$ to a second point $B$ and back again to $A$ after reflection at $B$. From elementary considerations one can see that its inverse is the average between the inverses of $c(\theta)$ and $c(\theta + \pi)$. A look at (2) then shows that it is given by

$$\frac{1}{c_2(\theta)} = \frac{R}{c}\left[\frac{\cos^2\theta}{f_1^2(1-\beta^2)^2} + \frac{\sin^2\theta}{g_2^2(1-\beta^2)}\right]^{1/2} . \tag{4}$$

There are two well-established empirical facts that can be used to restrict the generality of the transformations (1):

1) Constancy of the two-way velocity of light. The condition

$$c_2(\theta) = c \tag{5}$$

agrees with measurements of increasing precision and holds for all $\theta$ [8].

2) Slowing down of clocks at rest in $S$ with respect to those in $S_0$. A clock in the origin of $S$ satisfies $x_0 = \upsilon t_0$, and the last Eq. (1) is required to imply for the time $t$ it marks

$$t = \sqrt{1-\beta^2}\, t_0 . \tag{6}$$

It is easy to see that (5) and (6) can be satisfied if and only if

$$f_1 = \frac{1}{\sqrt{1-\beta^2}} \quad ; \qquad g_2 = 1 \quad ; \qquad e_4 = \sqrt{1-\beta^2} - e_1\,\upsilon , \tag{7}$$

so that in place of (1) one can write

$$\begin{cases} x = \dfrac{x_0 - \upsilon t_0}{\sqrt{1-\beta^2}} \\ y = y_0 \\ z = z_0 \\ t = \sqrt{1-\beta^2}\, t_0 + e_1(x_0 - \upsilon t_0) \end{cases} . \tag{8}$$

The only remaining unknown function of $\upsilon$ is $e_1$. This is a conventional term, sometimes called the "clock synchronization factor." Length contraction of rods moving with respect to $S_0$ by the usual factor $\sqrt{1-\beta^2}$ (independently of $e_1$) is also a consequence of (8). The velocity of light compatible with (8) can be obtained from (2) and (7) and turns out to be given by

$$\frac{1}{c(\theta)} = \frac{1 + \Gamma\cos\theta}{c} , \tag{9}$$

where

$$\Gamma = \beta + c e_1 \sqrt{1 - \beta^2} . \qquad (10)$$

The transformations (8) represent the complete set of theories "equivalent" to TSR: if $e_1$ is varied, different elements of this set are obtained, which, according to the conventionality thesis of Reichenbach [1-2], should all be equivalent as far as the explanation of experiments is concerned. The Lorentz transformation is recovered as a particular case with $e_1 = -\beta / c \sqrt{1-\beta^2}$ , whence it also follows that $c(\theta) = c$ . Different values of $e_1$ are obtained from different synchronization conventions. In all cases but that of TSR, such values imply the existence of a privileged frame [7]. In spite of this it can be shown that many basic experiments of relativistic physics are explained equally well by any theory of the set.

## 3. Length contraction and clock retardation

In this section we will generalize the Mansouri-Sexl result [3] that no physically meaningful difference from the TSR is to be expected from observations of length contraction and clock retardation effects, if a different synchronization is adopted. Their proof was limited to the equivalence of the cases $e_1 = 0$ (absolute synchronization) and $e_1 = -\beta / c \sqrt{1-\beta^2}$ (Einstein synchronization).

Consider a rod at rest on the $x$-axis of the inertial frame $S$ between the points $x_1$ and $x_2$. Let observers in $S_0$ mark the end points of the rod on their $x_0$ axis at the same time $t_0$, and let $x_{01}$ and $x_{02}$ be the coordinates so found. From (8) one gets

$$x_1 = \frac{x_{01} - \upsilon t_0}{\sqrt{1-\beta^2}} \quad ; \quad x_2 = \frac{x_{02} - \upsilon t_0}{\sqrt{1-\beta^2}} \qquad (11)$$

whence

$$x_{02} - x_{01} = \sqrt{1 - \beta^2} \left( x_2 - x_1 \right) \qquad (12)$$

Eq. (12) shows that a rod at rest in $S$ appears to be contracted when seen from $S_0$.

To see what happens if instead the rod is at rest in $S_0$ and is observed from $S$, we need the inverse transformations, which are easily found from (8):

$$\begin{cases} x_0 = \left(\sqrt{1-\beta^2} - e_1\, \upsilon\right)x + \dfrac{\upsilon t}{\sqrt{1-\beta^2}} \\ y_0 = y \\ z_0 = z \\ t_0 = \dfrac{1}{\sqrt{1-\beta^2}}\, t - e_1\, x \end{cases} \tag{13}$$

Consider now a rod at rest on the $x_0$ axis of the inertial frame $S_0$ between the points $x_{01}$ and $x_{02}$. Let two observers in $S$ mark the end points of the rod on their $x$ axis at the same time $t$, and let $x_1$ and $x_2$ be the coordinates so found (the symbols are the same as in the previous case, but of course their meaning is different). From the first Eq. (13) it follows that

$$x_{01} = \left(\sqrt{1-\beta^2} - e_1\, \upsilon\right)x_1 + \frac{\upsilon t}{\sqrt{1-\beta^2}};\; x_{02} = \left(\sqrt{1-\beta^2} - e_1\, \upsilon\right)x_2 + \frac{\upsilon t}{\sqrt{1-\beta^2}}, \tag{14}$$

whence

$$x_2 - x_1 = \frac{x_{02} - x_{01}}{\sqrt{1 - \beta^2} - e_1\upsilon}. \tag{15}$$

In this case the length depends on the synchronization parameter $e_1$, but this is not surprising because the very definition of the length of a moving rod is based on simultaneity: we said that two observers in $S$ mark the end points of the rod on their $x$ axis *at the same time* $t$. For example, if one takes $e_1 = -\beta/c\,\sqrt{1-\beta^2}$ in (15), as in the TSR, one gets $x_2 - x_1 = \sqrt{1-\beta^2}\left(x_{02} - x_{01}\right)$, the usual relativistic length contraction of moving rods. If in (15) one takes instead $e_1 = 0$ (absolute synchronization), one clearly gets the opposite effect, a lengthening of the rod.

Next consider a clock $W$ at rest on the $x$-axis of the inertial frame $S$ at the point $x_W$. Consider two observers at rest in $S_0$ at the points $x_{01}$ and $x_{02}$ and let them check the times $t_{01}$ and $t_{02}$ shown by their clocks when $W$ passes near to them, showing respectively the times $t_1$ and $t_2$. Since $W$ is at rest in $S$, it moves with velocity $\upsilon$ with respect to $S_0$ and one must have

$$x_{02} - x_{01} = \upsilon\left(t_{02} - t_{01}\right). \tag{16}$$

From the fourth Eq. (8) it follows that

$$\begin{aligned} t_1 &= \sqrt{1-\beta^2}\, t_{01} + e_1(x_{01} - \upsilon t_{01}) \\ t_2 &= \sqrt{1-\beta^2}\, t_{02} + e_1(x_{02} - \upsilon t_{02}) \end{aligned}, \tag{17}$$

whence, subtracting and taking into account (16):

$$t_{02} - t_{01} = \frac{t_2 - t_1}{\sqrt{1-\beta^2}}. \tag{18}$$

Seen from $S_0$ the clock at rest in $S$ appears to be retarded, because in (18) $t_2 - t_1$ must be smaller than $t_{02} - t_{01}$ for any $\beta \neq 0$.

Next consider a clock $W_0$ at rest on the $x_0$ axis of the inertial frame $S_0$ in the point $x_{0W}$. Consider two observers at rest in $S$ at the points $x_1$ and $x_2$ and let them check the times $t_1$ and $t_2$ shown by their clocks when $W_0$ passes near to them, showing the times $t_{01}$ and $t_{02}$, respectively. Since $W_0$ is at rest in $S_0$, it must satisfy the first Eq. (13), that is

$$x_{0W} = \left(\sqrt{1-\beta^2} - e_1\upsilon\right)x_1 + \frac{\upsilon t_1}{\sqrt{1-\beta^2}}\ ;\ x_{0W} = \left(\sqrt{1-\beta^2} - e_1\upsilon\right)x_2 + \frac{\upsilon t_2}{\sqrt{1-\beta^2}}, \tag{19}$$

whence, by subtraction

$$x_2 - x_1 = \frac{-\upsilon(t_2 - t_1)}{\sqrt{1-\beta^2}\left(\sqrt{1-\beta^2} - e_1\upsilon\right)}, \tag{20}$$

which describes the motion of $W_0$ relative to $S$. From the 4th Eq. (13) it follows that

$$t_{01} = \frac{1}{\sqrt{1-\beta^2}}\, t_1 - e_1 x_1\ ;\quad t_{02} = \frac{1}{\sqrt{1-\beta^2}}\, t_2 - e_1 x_2, \tag{21}$$

whence, by subtracting and taking (20) into account

$$t_2 - t_1 = \left(\sqrt{1-\beta^2} - e_1\upsilon\right)(t_{02} - t_{01}). \tag{22}$$

In this case the time difference depends on the synchronization parameter $e_1$. Again, this should not be considered surprising because to check the behaviour of a clock moving with respect to $S$ one must *in all cases* to compare it with two clocks at rest at different positions in $S$: thus the result depends on the synchronization convention adopted in $S$. For example, if in (22) one

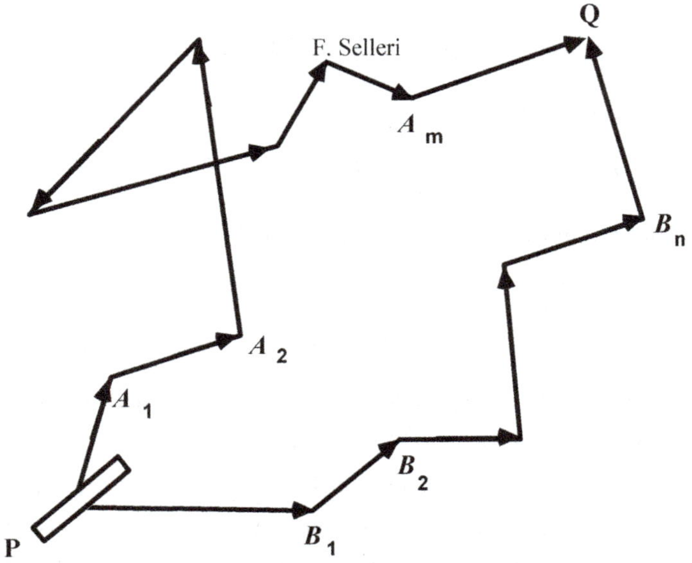


**Figure 2**. A semitransparent mirror placed in $P$ generates two coherent beams of light, which follow two different paths until they meet again in point $Q$ where they interfere.

takes $e_1 = -\beta / c \sqrt{1-\beta^2}$ , one gets $t_2 - t_1 = (t_{02} - t_{01}) / \sqrt{1-\beta^2}$ , the usual lower rate effect of the TRS. If in (22) one takes instead $e_1 = 0$ ("absolute synchronization"), one clearly gets the opposite effect, a faster rate of the clock at rest in $S_0$.

We conclude that no physically meaningful differences between theories based on different values of the parameter $e_1$ exist, insofar as one considers length measurements of moving rods, or time measurements of moving clocks. In fact, the observable differences depend entirely on the adopted clock synchronization convention, and have no physically objective basis.

## 4. Michelson-type Experiments

Consider a laboratory at rest in the inertial system $S$ moving with velocity $\vec{v}$ relative to the isotropic system $S_0$. In this laboratory an interferometric experiment is performed in which a beam of light is split into two parts by a semitransparent mirror placed in point $P$. The first part propagates along the path $P - A_1 - A_2 ... A_m - Q$, where a reflecting mirror is placed at every intermediate point, the second part along the similar path $P - B_1 - B_2 ... B_n - Q$, also equipped with mirrors. Finally the two parts superimpose at $Q$ where they interfere, $Q$ being an arbitrary point of an interference figure [9]. On the first path we define the vectors $\vec{\ell}_{ai}$ (with moduli $\ell_{ai}$), $i = 1, 2, ... m+1$, coincident with the rectilinear segments described by light and all oriented from $P$ toward $Q$; on the second path we similarly define the vectors $\vec{\ell}_{bj}$ (with moduli $\ell_{bj}$), $j = 1, 2, \ldots n+1$.

The interference in $Q$ is determined by the time delay $\Delta T$ between the two rays. The prediction of the TSR is easy to find. In this theory light propagates in all directions with the same speed $c$, also with respect to $S$, and one has

$$\Delta T = T_B - T_A = \frac{L_B - L_A}{c}, \tag{23}$$

where

$$L_A = \sum_{i=1}^{m+1} \ell_{ai} \quad ; L_B = \sum_{j=1}^{n+1} \ell_{bj}. \tag{24}$$

Next we calculate $\Delta T$ from the equivalent transformations, according to which the inverse velocity of light relative to $S$ is given by (9). The time delay is now given by

$$\Delta T = \sum_{j=1}^{n+1} \frac{\ell_{bj}}{c(\theta_{bj})} - \sum_{i=1}^{m+1} \frac{\ell_{ai}}{c(\theta_{ai})}, \tag{25}$$

where $\theta_{ai}$ ($\theta_{bj}$) is the angle between $\vec{\ell}_{ai}$ and $\vec{\upsilon}$ ($\vec{\ell}_{bj}$ and $\vec{\upsilon}$). By inserting (9) in (25) one has

$$\begin{aligned} \Delta T &= \frac{L_B - L_A}{c} + \frac{\Gamma}{c}\sum_{j=1}^{n+1} \ell_{bj}\cos\theta_{bj} - \frac{\Gamma}{c}\sum_{i=1}^{m+1} \ell_{ai}\cos\theta_{ai} \\ &= \frac{L_B - L_A}{c} + \frac{\Gamma}{c}\left[\sum_{j=1}^{n+1} \vec{\ell}_{bj} - \sum_{i=1}^{m+1} \vec{\ell}_{ai}\right]\cdot\frac{\vec{\upsilon}}{\upsilon} \\ &= \frac{L_B - L_A}{c} \end{aligned} \tag{26}$$

The last step is a consequence of

$$\sum_{j=1}^{n+1} \vec{\ell}_{bj} = \sum_{i=1}^{m+1} \vec{\ell}_{ai}, \tag{27}$$

and (27) holds because the two sides are separately equal to the vector joining $P$ and $Q$. The results (23) and (26) are the same. Therefore $\Gamma$ (containing $e_1$) disappears from the result and all theories based on the equivalent trans-

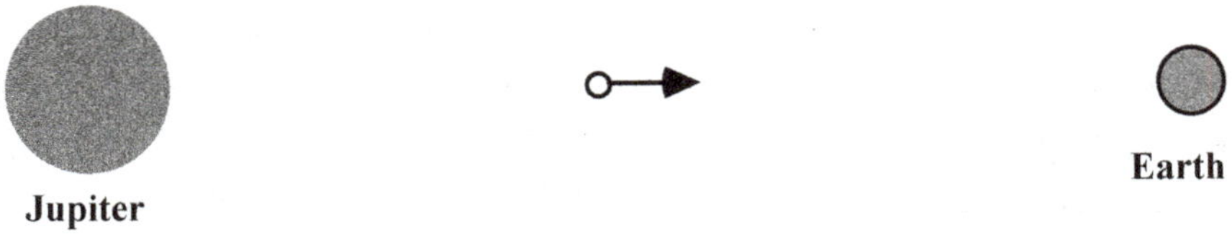


**Figure 3**. A light signal is emitted by one of Jupiter's satellites and later detected on the Earth as an occultation.

formations lead to the same predictions for interferometric experiments of the Michelson type (Michelson-Morley [10], Kennedy-Thorndike [11], Q. Majorana [12], *etc.*).

## 5. Occultations of Jupiter's Satellites

Let a satellite of Jupiter (*e.g.*, Io) be in any state of motion on the $x_0$ axis of the isotropic inertial system $S_0$ dealt with in Section 2. Io sends a light signal (occultation) at the $S_0$ time $T_0$ when it is in the position

$$x_{0I} = -L_0 \tag{28}$$

The equation of motion of the signal relative to $S_0$ is

$$x_0 = -L_0 + c\left(t_0 - T_0\right); \tag{29}$$

$\left(t_0 \geq T_0\right)$. The Earth moves with constant speed $v$ and constitutes instantaneously the inertial system $S$ of Fig. 1, where it is taken to be in the origin of the system of Cartesian axes.

The equation of motion of Earth as seen from $S_0$ is

$$x_{0E} = \upsilon t_0 . \tag{30}$$

The signal arrives on Earth at time $t_{0a}$, which, because of (29)-(30) must satisfy

$$\upsilon t_{0a} = -L_0 + c\left(t_{0a} - T_0\right), \tag{31}$$

whence

$$t_{0a} = \frac{L_0 + cT_0}{c - \upsilon}. \tag{32}$$

The Earth position at the time of arrival is

$$x_{0a} = \upsilon t_{0a} \tag{33}$$

Our present problem is to find the time $t_a$ marked by a clock on Earth when the signal is received. Between $S_0$ and $S$, the transformations (1) apply and one must have

$$t_a = e_1 x_{0a} + e_4 t_{0a} = (e_1 \upsilon + e_4) t_{0a}, \tag{34}$$

whence, using (32)

$$t_a = (e_1 \upsilon + e_4) \frac{L_0 + c T_0}{c - \upsilon}. \tag{35}$$

But in all theories equivalent to the TSR the last Eq. (7) holds—the time dilation condition for the $S$ clocks with respect to clocks in $S_0$. Therefore, in place of (35) one must write

$$t_r = \sqrt{1 - \upsilon^2/c^2} \, \frac{L_0 + c T_0}{c - \upsilon}. \tag{36}$$

As can be seen, on the right hand side all the quantities $\upsilon$, $c$, $L_0$, $T_0$ are measured in $S_0$ and, therefore, have values that are exactly the same in all equivalent theories, independently of $e_1$ and of how clocks have been synchronized in $S$. Hence, all the equivalent theories predict exactly the same occultation time to be observed on Earth.

## 6. Aberration

The following notation will be used:

$\vec{\upsilon}$ is the velocity of Earth with respect to the fundamental frame $S_0$ ("absolute velocity," variable during the year).

$\vec{c}$ is the vector velocity of a light ray with respect to $S_0$. Its modulus has the constant value $c$.

$\vec{c}_\upsilon$ is the velocity of the same light-ray relative to Earth.

We start by reviewing the classical treatment of the problem. From Fig. 4 it follows that

$$\vec{c}_r = \vec{c} - \vec{\upsilon}, \tag{37}$$

which is the same as

$$-c_r \cos\theta = -c \cos\theta_0 - \upsilon \ ; \qquad -c_r \sin\theta = -c \sin\theta_0, \tag{38}$$

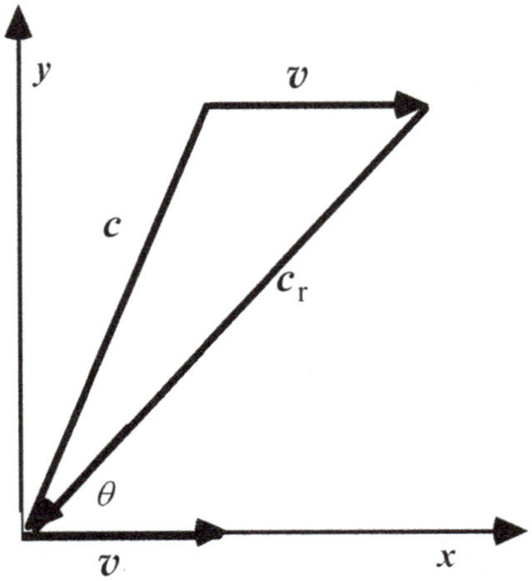


**Figure 4**. Kinematic quantities for the aberration of light as seen from the fundamental reference frame.

whence

$$\tan\theta = \frac{c \sin\theta_0}{c \cos\theta_0 + \upsilon}. \tag{39}$$

From Eq. (39) one sees that the difference between $\theta_0$ and $\theta$ is very small (of the order of $\upsilon/c$). The usual (approximate) aberration formula is deducible from (39).

We come next to the general solution of the aberration problem. The generalized transformations (1) from the privileged frame to the Earth frame in two dimensions and in differential form are

$$\begin{cases} d x_E = f_1 \left[d x_0 - \upsilon d t_0\right] \\ d y_E = g_2 d y_0 \\ d t_E = e_1 d x_0 + e_4 d t_0 \end{cases}, \tag{40}$$

where the index "$E$" denotes quantities calculated with respect to the inertial frame in which the Earth is instantaneously at rest. Consider an object of any nature and define its velocity components with respect to the Earth frame and to the isotropic frame $S_0$:

$$u_{Ex} = \frac{d x_E}{d t_E} \;;\; u_{Ey} = \frac{d y_E}{d t_E} \;;\; u_{0x} = \frac{d x_0}{d t_0} \;;\; u_{0y} = \frac{d y_0}{d t_0}. \tag{41}$$

Dividing side by side the first two Eqs. (40) by the third one, it follows that

$$u_{Ex} = \frac{f_1\left(u_{0x} - \upsilon\right)}{e_1 u_{0x} + e_4} \quad ; \quad u_{Ey} = \frac{g_2 u_{0y}}{e_1 u_{0x} + e_4}. \tag{42}$$

Referring now to the propagation of the light pulse of Fig. 4, and writing

$$\begin{aligned} u_{Ex} &= -c_r \cos\theta \quad ; \quad u_{Ey} = -c_r \sin\theta \\ u_{0x} &= -c \cos\theta_0 \quad ; \quad u_{0y} = -c \sin\theta_0, \end{aligned} \tag{43}$$

we get from (42)

$$c_r \cos\theta = \frac{f_1\left(c\cos\theta_0 + \upsilon\right)}{e_1 u_{0x} + e_4} \quad ; \quad c_r \sin\theta = \frac{g_2 c \sin\theta_0}{e_1 u_{0x} + e_4}, \tag{44}$$

which is the generalization of (38). Dividing side by side the two Eqs. (44) we obtain

$$\tan\theta = \frac{g_2}{f_1} \frac{c\sin\theta_0}{c\cos\theta_0 + \upsilon} = \sqrt{1 - \beta^2}\, \frac{\sin\theta_0}{\cos\theta_0 + \beta}, \tag{45}$$

the last step being a consequence of (7). Thus, we see that the difference between (45) and (39) is only due to terms of the second and higher order in $\beta$. Clearly all consequences of the classical theory hold also in the general case. Therefore (45) is in agreement with the experimental evidence [13].

The really important point, however, is the following. All terms appearing in the right hand side of (45) ($\upsilon$, $c$, $\theta_0$) are measured in the fundamental frame. Given that what is observed in that frame is the same in all equivalent theories, we also see that the angle $\theta$ perceived on Earth is predicted to be the same. Thus, the variations of $\theta$ are predicted to be exactly the same by all theories equivalent to TSR [14].

## 7. Radar Ranging of Planets

Let the equations of motion, written in the fundamental inertial frame $S_0$, of Earth, of Venus, and of a radar signal sent from Earth toward Venus respectively be

$$x_{01} = \upsilon t_0; \; x_{02} = \upsilon_2 t_0 + d_0; \quad x_0 = c t_0. \tag{46}$$

The time $t_{0R}$ at which the reflection of the radar pulse on the surface of Venus takes place [15] must, therefore, satisfy the condition

$$c\,t_{0R} = \upsilon_2\, t_{0R} + d_0 \qquad \Rightarrow \qquad t_{0R} = \frac{d_0}{c - \upsilon_2}. \tag{47}$$

During the return journey of the radar pulse from Venus to Earth the latter still obeys the first Eq. (46), while the pulse must satisfy

$$x_0 = x_{0R} - c\,(t_0 - t_{0R}). \tag{48}$$

In (48) we introduced

$$x_{0R} = c\,t_{0R} = \frac{d_0\, c}{c - \upsilon_2}, \tag{49}$$

the position occupied jointly by Venus and the pulse at time $t_{0R}$. Therefore the arrival time $t_{0A}$ of the pulse on Earth must satisfy

$$\upsilon\, t_{0A} = x_{0R} - c\,(t_{0A} - t_{0R}). \tag{50}$$

Using (47) and (49), the previous equation gives

$$t_{0A} = \frac{2\,d_0 c}{(c - \upsilon_2)(c + \upsilon)}. \tag{51}$$

From $S_0$ to $S$ (the Earth rest frame) the transformations (1) apply. Obviously, for the pulse arrival time measured on Earth, one must have

$$t_A = e_1\, x_{0A} + e_4\, t_{0A} = \left(e_1\, \upsilon + e_4\right) t_{0A}, \tag{52}$$

where $x_{0A} = \upsilon\, t_{0A}$ is the Earth's position at the time $t_{0A}$ when the radar signal is received. From (51) and the third Eq. (7) it follows that

$$t_A = \sqrt{1 - \upsilon^2 / c^2}\, \frac{2\,d_0 c}{(c - \upsilon_2)(c + \upsilon)}. \tag{53}$$

As can be seen, on the right hand side all the quantities $c$, $d_0$, $\upsilon$, $\upsilon_2$ are measured in $S_0$ and, therefore, have values which are exactly the same in all equivalent theories, independently of clock synchronization in $S$ and thus of $e_1$. Thus, all such theories predict that the same arrival time of the radar pulse will be observed on Earth.

## 8. Conclusions

The previous results agree with those obtained by Mansouri and Sexl [3], Will [4], and Croca and Selleri [16]. Experimental tests of the isotropy of the speed of light using one way propagations [17]-[21] were analysed by Will [4], who found that, when properly expressed in terms of measurable quantities, the results of these experiments are independent of the clock synchronization method. This conclusion is in agreement with the conclusions reached in the present paper, which shows that the isotropy of the speed of light relative to moving systems has never been checked experimentally. Our approach differs from Mansouri and Sexl [3] and Will [4] in that these authors worked with what they called a "test theory of special relativity" [equivalent to our transformations (1)] in an approximate way, while we have taken the well-established experimental points expressed by Eqs. (5) and (6) into account and made rigorous calculations starting from the transformations (8). The case for an empirical equivalence of all transformations of the type (8), independently of $e_1$, is strengthened by our results The equivalence holds for experiments carried out in inertial frames. It is important to stress that a natural requirement of physical continuity between inertial systems and systems possessing a small acceleration leads to a breakdown of equivalence, favouring the choice of $e_1 = 0$ [22].

The kinematics of high-energy particle interactions has been studied [23], showing that complete equivalence exists between the theory with $e_1 = 0$ and the TSR. Therefore, the kinematics of high energy collisions, the determination of particle masses, and so on, do not require a different analysis from the one successfully carried out up to the present time. When $e_1 = 0$, the formulae for energy and momentum have the same mathematical expression in SRT and in the theory based on inertial transformations (ITT), but only relative to the fundamental isotropic frame. Relative to other inertial systems they are mathematically different, although numerically equal, this being possible because the same symbol has different values in the two theories (*e.g.*, a relative velocity). The coincidence is only numerical (and not also analytical) because the dependence on the one-way velocity of the particle is different in the two theories.

## References

[1] H. Reichenbach, *The Philosophy of Space and Time*, Dover, New York (1930).

[2] M. Jammer, "Some fundamental problems in the special theory of relativity," in *Problems in the Foundations of Physics*, G. Toraldo di Francia ed., p. 202, Società Italiana di Fisica, Bologna, and North Holland, Amsterdam (1979).

[3] R. Mansouri, R. Sexl, *General Relat. Gravit.* **8**, 497, 515, 809 (1977).

[4] C.M. Will, *Phys. Rev.* D **45**, 403 (1992).

[5] H. Poincaré, *Rev. Metaphys. Morale* **6**, 1 (1898).

[6] A.Einstein, *Relativity, The Special, The General Theory*, Chicago (1951). See explicit statement on p. 18.

[7] F. Selleri, *Chin. J. Eng. Elect.* **6**, 25 (1995); *Found. Phys.* **26**, 641 (1996).

[8] Presently the two-way velocity of light in vacuum is known with an error of about 20 cm/s and has the value $c = (299\ 792\ 458.8 \pm 0.2)$ m/s. See: P.T. Woods, K.C. Sholton, W.R.C. Rowley, *Appl. Opt.* **17**, 1048 (1978); D.A. Jennings, R.E. Drullinger, K.M. Evenson, C.R. Pollock, J.S Wells, *J. Res. Natl. Bur. Stand.* **92**, 11 (1987).

[9] A similar proof was given in [7], but was limited to the case $e_1 = 0$.

[10] A.A. Michelson and E.W. Morley, *Am. J. Science*, **34**, 333 (1887).

[11] R.J. Kennedy and E.M.Thorndike, *Phys.Rev.* **42**, 400-418 (1932).

[12] Q. Majorana, *Phil. Mag.* **37**, 145-150 (1919).

[13] *E.g.,* see: T.E. Phipps, Jr., *Am. J. Phys.* **57**, 549 (1989).

[14] For aberration see: F. Goy, *Found. Phys. Lett.* **9**, 165 (1996).

[15] S.J. Ostro, *Rev. Mod. Phys.* **65**, 1235 (1993).

[16] J.R. Croca and F. Selleri, *Nuovo Cim.* **114 B**, 447 (1999).

[17] E. Riis, L.-U. Anderson, N. Bjerre, O. Poulsen, S.A. Lee, and J.L. Hall, *Phys. Rev. Lett.* **60**, 81 (1988); **62**, 842 (1989).

[18] T.P. Krisher, L. Malecki, G.F. Lutes, L.E. Primas, R.T. Logan, J.D. Anderson, and C.M. Will, *Phys. Rev.* D **42**, 731 (1990).

[19] K.C. Turner and H.A. Hill, *Phys. Rev.* **134**, B252 (1964).

[20] D.C. Champeney, G.R. Isaak, and A.M. Kahn, *Phys. Lett.* **7**, 241 (1970).

[21] R.F.C. Vessot, M.W. Levine, E.M. Mattison, E.L. Blomberg, T.E. Hoffman, G.U. Nystrom, B.F. Farrell, R. Decher, P.B. Eby, C.R. Baugher, J.W. Watts, D.L. Teuber, and F.O. Wills, *Gen Relativ. Gravit.* **10**, 181 (1979).

[22] F. Selleri, *Space and Time should be preferred to Spacetime - II*, second paper presented at this conference.

[23] F. Selleri, *Found. Phys. Lett.* **9**, 43 (1996).

# Space and Time should be Preferred to Spacetime - 2

F. Selleri[8]

A physical quantity $\rho$ exists for which the theory of special relativity (TSR) predicts $\rho = 1$ relative to all inertial frames. Under extremely general conditions we prove that $\rho = (c+\upsilon)/(c-\upsilon)$ for all rotating disks having the same peripheral velocity $\upsilon$ and *arbitrarily small acceleration* $a$. This value of $\rho$ must hold in any small region near the disk rim. Therefore, the TSR gives rise to a discontinuity. The limit $a \to 0$ should instead be smooth, because all empirical knowledge about inertial systems is obtained in frames with $a \neq 0$, *e.g.*, because of the Earth's rotation. Elimination of the discontinuity is possible using the set of theories "equivalent" to TSR of Part I. The clock synchronization ambiguity in inertial systems is then solved: only $e_1 = 0$ (corresponding to absolute simultaneity) gives $\rho = (c+\upsilon)/(c-\upsilon)$ when $a \to 0$. Non-invariant values of the one-way velocity of light are thus obtained.

## 1. Time on Rotating Platforms

The basic idea of this paper is that inertial frames can always be treated as particular cases of accelerated frames with zero acceleration. The limit $a \to 0$, however, is not really needed from the physical point of view. No perfectly inertial frame exists in practice, *e.g.*, because of the Earth's rotation around its axis, orbital motion, *etc.* All we know about inertial systems has actually been obtained in frames having a small, but non-zero, acceleration. Of course *in the theoretical schemes* the mathematical limit $a \to 0$ can be taken, and it must be a smooth limit, without discontinuities in $\rho$ between slowly accelerated systems and inertial systems. Otherwise, the physical reality would contradict the theory of inertial systems. From this point of view, the TSR will be seen to be unsatisfactory.

Consider an inertial reference system $S_0$ and assume it to be isotropic: the one-way velocity of light relative to $S_0$ has the usual value $c$ in all direc-

[8] Università di Bari - Dipartimento di Fisica, INFN - Sezione di Bari, Via Amendola 173, I-70126 Bari, Italy

tions. In relativity the latter assumption is true in all inertial frames, while in other theories only one such frame exists [1]. A laboratory in which physical experiments are performed is assumed to be at rest in $S_0$, and clocks in it to be synchronized with the Einstein method, that is, by means of light signals.

In this laboratory there is a perfectly circular platform having radius $R$ which rotates around its axis with angular velocity $\omega$ and peripheral velocity $\upsilon = \omega R$. On the platform rim there is a single clock $C_\Sigma$ showing the time $t$. Unicity of $C_\Sigma$ makes sure that synchronization problems do not arise on the platform. We assume $C_\Sigma$ to be set as follows: When a clock of the laboratory momentarily very near $C_\Sigma$ shows time $t_0 = 0$, then $C_\Sigma$ is also set at time $t = 0$. Furthermore, if the platform were not rotating, $C_\Sigma$ would always show the same time $t_0$ as the laboratory clocks. When it does rotate, however, its motion changes the rate of $C_\Sigma$. The relationship between $t$ and $t_0$ is taken to have the form

$$t_0 = tF(\upsilon, a), \tag{1}$$

where $F$ is a function of velocity $\upsilon$, acceleration $a = \upsilon^2/R$, and eventually of higher derivatives of position (not shown).

We are, of course, far from ignorant about the function $F$. In the limit of small acceleration and constant velocity it is expected to become:

$$F(\upsilon, 0) = \frac{1}{\sqrt{1-\upsilon^2/c^2}}. \tag{27}$$

There are even strong indications that the dependence on acceleration in (1) is absent in all cases [2]. This is, however, unimportant for our present purposes, because the results obtained below hold for all possible functions $F(\upsilon, a)$.

## 2. Velocity of light on rotating platforms

On the rim of the platform, besides $C_\Sigma$ there is a light source $\Sigma$ placed at a fixed position very near $C_\Sigma$. Two oppositely moving light flashes leave $\Sigma$ at time $t_1$ of $C_\Sigma$ and are forced to move circularly by "sliding" on the internal surface of a cylindrical mirror placed at rest on the platform all around it and very near its border. If the mirror is ignored, the light flashes propagate in the vacuum. The motion of the mirror cannot change the velocity of light, because the mirror is like a source (a "virtual" one) and the motion of a source never changes the velocity of the light signals originating from it. Thus, relative to the laboratory, the light flashes propagate with the usual velocity $c$.

The description of light propagation given by the $S_0$ (laboratory) observers is the following: Two light flashes leave $\Sigma$ at time $t_{01}$. The first one propagates circularly in the sense opposite to the platform's rotation and comes back to $\Sigma$ at time $t_{02}$ after a full rotation around the platform. The second one propagates circularly in the direction of the platform's rotation and return to $\Sigma$ at time $t_{03}$ after a full rotation around the platform. These laboratory times, all relative to events taking place at points of the platform very near $C_\Sigma$, are related to the corresponding $C_\Sigma$ times *via* Eq. (1)

$$t_{0i} = t_i \, F(\upsilon, a) \qquad (i=1,2,3). \tag{3}$$

Light propagating in the direction opposite to the disk rotation must cover a distance smaller than the disk circumference length $L_0$ by a quantity $x = \omega R\left(t_{02} - t_{01}\right)$ equalling the shift of $\Sigma$ during the time $t_{02} - t_{01}$ taken by light to reach $\Sigma$. Therefore

$$L_0 - x = c(t_{02} - t_{01}) \;\; ; \quad x = \omega R(t_{02} - t_{01}) \, . \tag{4}$$

It follows that

$$t_{02} - t_{01} = \frac{L_0}{c + \upsilon} . \tag{5}$$

Light propagating in the rotational direction of the disk must instead cover a distance larger than $L_0$ by a quantity $y = \omega R(t_{03} - t_{01})$ equalling the shift of $\Sigma$ during the time $t_{03} - t_{01}$ taken by light to reach $\Sigma$. Therefore

$$L_0 + y = c\,(t_{03} - t_{01}) \quad ; \quad y = \omega R(t_{03} - t_{01}). \tag{6}$$

One now gets

$$t_{03} - t_{01} = \frac{L_0}{c - \upsilon} . \tag{7}$$

From (5) and (7) one gets

$$\frac{t_{03} - t_{01}}{t_{02} - t_{01}} = \frac{c + \upsilon}{c - \upsilon} . \tag{8}$$

We show next that these laboratory relations fix to some extent the velocity of light relative to the disk. In fact, (3) applied to (8) gives

$$\frac{t_3 - t_1}{t_2 - t_1} = \frac{c + \upsilon}{c - \upsilon}, \tag{9}$$

where, of course, $t_2 - t_1$ and $t_3 - t_1$ are the propagation times for the light pulses to describe a complete rotation along the platform rim in the two directions, as measured by $C_\Sigma$. But the propagation times are inversely proportional to the velocities [3]. Therefore, if $c_g(0)$ and $c_g(\pi)$ are the global light velocities, relative to the disk, for the flash propagating along the disk border in the direction of rotation, and in the opposite direction, respectively, from (9) it follows that one must have

$$\frac{c_g(\pi)}{c_g(0)} = \frac{c + \upsilon}{c - \upsilon}. \tag{10}$$

Notice that $F(\upsilon, a)$ has disappeared from the ratios (9) and (10). The result (10) has been deduced under extremely general conditions: in practice, only space isotropy in the laboratory (inertial) frame was assumed.

As we will see, the implication that $c_g(0) \neq c_g(\pi)$ is a big problem for the TSR: instantaneous light velocities relative to the disk cannot be equal if global velocities are different! By virtue of a "continuity principle" between slowly accelerated and inertial systems, we will show that (10) must also hold for light velocities in "comoving" inertial frames [4].

## 3. Instantaneous light velocity on rotating platforms

One can easily see that (10) gives not only the ratio of the global light velocities for full trips around the platform in opposite directions, but the ratio of instantaneous light velocities in any small region of the disk rim as well. Isotropy of space ensures that the instantaneous velocity of light in the direction concordant with the disk rotation is the same at all points of the rim, so that it has to coincide with the global velocity. The same holds in the discordant direction. This argument is spelled out in greater detail below.

Let the disk circumference be divided into a very large number $n$ of segments, all with length $\ell = 2\pi R/n$, if $R$ is the disk radius. In any reasonable theory the time $T$ taken by a light pulse to make a trip around the disk equals the sum of the times taken for all segments (no matter how they may be measured),

$$T = \sum_{i=1}^{n} t_i \,. \tag{11}$$

Given the circumference length $L = 2\pi R$ (measured on the disk), one can write (11) as

$$\frac{T}{L} = \sum_{i=1}^{n} \frac{\ell}{L} \frac{t_i}{\ell} \,. \tag{12}$$

The global velocity $c_g$ and the local velocities $\tilde{c}_i$ are defined as

$$c_g \equiv \frac{L}{T} \quad ; \quad \tilde{c}_i \equiv \frac{\ell}{t_i} , \tag{13}$$

so that (12) becomes

$$\frac{1}{c_g} = \frac{\ell}{L} \sum_{i=1}^{n} \frac{1}{\tilde{c}_i} \,. \tag{14}$$

The argument leading from (11) to (14) can be repeated for the two flashes of light going in opposite directions. Eq. (14) holds for both, that is, for propagation concordant $(\theta = 0)$ and discordant $(\theta = \pi)$ with disk rotation:

$$\frac{1}{c_g(0)} = \frac{\ell}{L} \sum_{i=1}^{n} \frac{1}{\tilde{c}_i(0)} ; \qquad \frac{1}{c_g(\pi)} = \frac{\ell}{L} \sum_{i=1}^{n} \frac{1}{\tilde{c}_i(\pi)} \,. \tag{15}$$

But (10) implies that $c_g(0) \neq c_g(\pi)$, and it then follows that it is impossible to satisfy the condition $\tilde{c}_i(0) = \tilde{c}_i(\pi)$ for all segments, no matter what synchronization is adopted for local clocks that may eventually be introduced. The introduction of local clocks is not necessary, however. Our argument has ontological value, and reality comes first, measurement only second!

From the point of view of the observers in the inertial system $S_0$ (assumed isotropic from the start) all points of the disk rim are physically equivalent, and it would be very strange if the rotating observers found local velocities of light different from one another. Therefore, the right synchronization on the disk is such that

$$\begin{cases} \tilde{c}_1(0) = \tilde{c}_2(0) \ldots = \tilde{c}_n(0) \equiv \tilde{c}(0) \\ \tilde{c}_1(\pi) = \tilde{c}_2(\pi) \ldots = \tilde{c}_n(\pi) \equiv \tilde{c}(\pi) \end{cases} , \tag{16}$$

where we have introduced the new symbols $\tilde{c}(0)$ and $\tilde{c}(\pi)$ to denote the local speeds of light in the two directions, now known to have the same values in all small regions of the disk rim. From Eq. (15) it then follows that

$$\tilde{c}(0) = c_g(0) \qquad ; \qquad \tilde{c}(\pi) = c_g(\pi) \tag{17}$$

We can therefore conclude that Eq. (10) gives not only the ratio of the global light velocities for full trips around the platform in opposite directions, but also the ratio of instantaneous velocities in any small region of the disk rim. In other words,

$$\frac{\tilde{c}(\pi)}{\tilde{c}(0)} = \frac{c+\upsilon}{c-\upsilon}. \tag{18}$$

We will see in the next section that this is difficulty cannot be overcome in a rational way by the TSR, which would require $\tilde{c}(0) = \tilde{c}(\pi) = c$ for all little segments of the disk rim.

## 4. Velocity of light relative to inertial frames

The motion of a rod and a clock placed on the platform rim can be considered almost rectilinear and uniform for a very short time. Such a rod and clock, if near to one another, are almost at rest in the same "tangent" (comoving) inertial system for time intervals that are short, but increase with disk radius. All equations valid on the rotating disk should transform smoothly into the corresponding equations of the inertial system. This is particularly true for the numerical values of velocities.

Consider $n$ platforms with radii $R_1, R_2, \ldots R_n$ ($R_1 < R_2 < \ldots < R_n$), and let them spin with angular velocities $\omega_1, \omega_2, \ldots \omega_n$ such that

$$\omega_1 R_1 = \omega_2 R_2 = \ldots \omega_n R_n = \upsilon .$$

Eq. (18) clearly applies, with the same $\upsilon$, to all of them. The accelerations

$$\frac{\upsilon^2}{R_1}, \frac{\upsilon^2}{R_2}, \ldots \frac{\upsilon^2}{R_n}$$

tend to zero for large $R_n$. Therefore, a small portion $\Pi$ of the rim of one of these platforms having very large $R_n$, must for a short time be locally equivalent to an inertial reference frame $S$ having the same instantaneous velocity relative to $S_0$, since $\Pi$ and $S$ have then not only the same $\upsilon$, but practically also the same (null) acceleration. It follows that Eq. (18) must apply to $S$ also.

This holds for all inertial frames, because any small region of any one of them can be imagined to be coincident with a portion of the border of a very large platform having instantaneously the same $\vec{\upsilon}$ relative to $S_0$ and $a \approx 0$.

The requirement that (18) applies to inertial reference systems is an application of what might be called *principle of continuity* between slowly accelerated systems and instantaneously comoving inertial systems. The idea can be formulated as follows:

> *The descriptions of the physical reality given in: (a) any small region $\Pi$ of a slowly accelerated system; (b) in the inertial system having the same instantaneous velocity as $\Pi$, should be very similar. They should become equal when the acceleration tends to zero.*

This principle becomes not only very natural, but conceptually unavoidable, if one considers the absence in nature of perfectly inertial systems. Such systems do not exist, because: (i) The earth is spinning; (ii) The earth has an orbital motion; (iii) The Milky Way is a rotating spiral galaxy. Therefore, no experiment has ever been performed in a truly inertial system. Nevertheless, we believe we know a lot about them, because we are convinced that very small accelerations have no practical effect.

One could say that our continuity principle is a weaker form of the idea introduced by Einstein in his 1911 paper on gravitation, where the red shift of light in a gravitational field was calculated, by means of the equivalence principle, as a Doppler effect due to relative motion of two inertial systems [5]. For this calculation Einstein did not even require that the system's acceleration was slow. The possible effects of acceleration were simply ignored.

The continuity principle can also be applied the other way around, from inertial to accelerated systems. This has been done in many books and papers, *e.g.* in the Landau-Lifschitz book [6] where an attempt is made to use the Einstein synchronization on the rotating platform, with the result that a clock cannot even be synchronized with itself. Our previous considerations should make it clear that a velocity of light equal to $c$ in opposite directions on the platform leads to $\rho = 1$, in contradiction with our basic theorem (18). Furthermore, if one sticks to the Einstein synchronization in inertial systems,

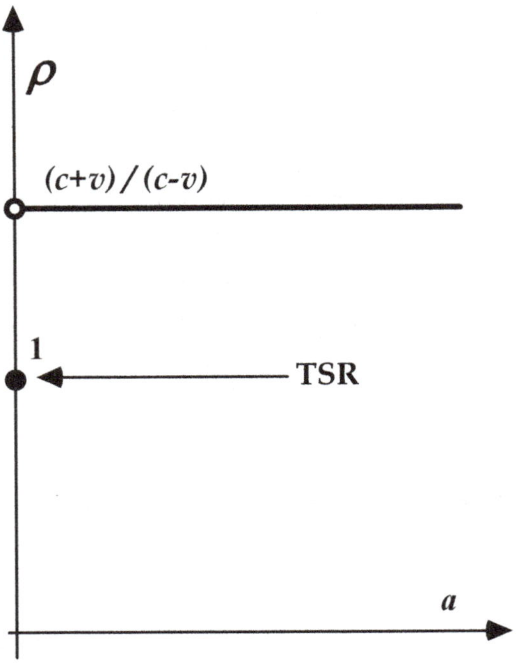


**Figure 1**. The ratio $\rho = \tilde{c}(\pi)/\tilde{c}(0)$ as a function of acceleration $a$ for rotating disks of constant peripheral velocity $\upsilon$ and increasing radius (decreasing $a$). The prediction of TSR for inertial systems ($a = 0$) is discontinuous.

the continuity step cannot be made and the logical situation becomes that of Fig. 1

We repeat that for all (moving) inertial systems a set of rotating platforms of the previous type can be found such that this result applies. In fact, any small region of any inertial system can be imagined coincident with a piece of the border of a very large platform having instantaneously the same $\vec{\upsilon}$ relative to $S_0$ and very small acceleration.

## 5. The right choice of synchronization

We have seen that the inverse speed of light compatible with the general set of space and time transformations between inertial systems given by Eq. (8) of Part I is:

$$\frac{1}{c(\theta)} = \frac{1}{c} + \left[\frac{\upsilon}{c^2} + e_1 R(\upsilon)\right]\cos\theta \ , \tag{19}$$

where

$$R(\upsilon) = \sqrt{1 - \upsilon^2/c^2},$$

and $\theta$ is the angle between the direction of propagation of light and the absolute velocity $\vec{\upsilon}$ of $S$. The transformations (8) of Part I represent the complete set of theories "equivalent" to the TSR: if $e_1$ is varied, different elements of this set are obtained. The Lorentz transformation is found as a particular case with $e_1 = -\upsilon/c^2 R(\upsilon)$. Different values of $e_1$ are obtained from different clock-synchronization conventions. In all cases but that of TSR, such values exclude the validity of at least the strong form of the relativity principle (the form used in deducing the Lorentz transformations), and imply the existence of a privileged frame. For all the theories represented by (19), only subluminal motions are possible ($\upsilon < c$).

In the previous sections we found a ratio of the one-way velocities of light along the rim of the disk, and relative to the disk itself, different from 1 and given by Eq. (18). Our principle of local continuity between the rim of the disk and the "tangent" inertial frame requires

$$\frac{c(\pi)}{c(0)} = \frac{c+\upsilon}{c-\upsilon}. \tag{20}$$

Eq. (19) applied to the cases $\theta = 0$ and $\theta = \pi$ becomes

$$\frac{1}{c(0)} = \frac{1}{c} + \left[\frac{\upsilon}{c^2} + e_1 R(\upsilon)\right]; \quad \frac{1}{c(\pi)} = \frac{1}{c} - \left[\frac{\upsilon}{c^2} + e_1 R(\upsilon)\right]. \tag{21}$$

This gives

$$\frac{c(\pi)}{c(0)} = \frac{c+\upsilon+e_1 R(\beta)c^2}{c-\upsilon-e_1 R(\beta)c^2}, \tag{22}$$

which clearly can agree with (20) if and only if

$$e_1 = 0. \tag{23}$$

The space-dependent term in the transformation of time is thus seen to disappear from the acceptable transformations. In this way, absolute simultaneity emerges as a necessary property of nature: two events taking place at different points and considered simultaneous by observers at rest in $\upsilon_0$ must also be judged simultaneous by the observers at rest in $S$.

## 6. The inertial transformations

In the previous section we showed that the condition $e_1 = 0$ must necessarily be used. This gives rise to the following transformation of space and time:

$$\begin{cases} x = \dfrac{x_0 - \beta c t_0}{R(\beta)} \\ y = y_0 \quad ; \quad z = z_0 \\ t = R(\beta) t_0 \end{cases} \tag{24}$$

where $\beta = \upsilon / c$. As already stressed by Mansouri and Sexl, these transformations would have been the logical consequence of a development along the line of thought of Lorentz-Larmor-Poincaré: they are the very relations one would write down if one had to formulate a theory in which rods shrink and clocks are slow by the usual factor when moving with respect to the ether. That the actual development followed a different line was due to the fact that "local time" was introduced at an early stage in considering the covariance of the Maxwell equations.

The one-way speed of light predicted by (24) can easily be found by taking $e_1 = 0$ in (19). It is

$$\frac{1}{c(\theta)} = \frac{1 + \beta \cos\theta}{c}. \tag{25}$$

The transformation (24) can be inverted and gives

$$\begin{cases} x_0 = R(\beta)\left[ x + \dfrac{\beta c}{R^2(\beta)} t \right] \\ y_0 = y \quad ; \quad z_0 = z \, . \\ t_0 = \dfrac{1}{R(\beta)} t \end{cases} \tag{26}$$

Note the formal difference between (24) and (26). The latter implies, for example, that the origin of $S_0$ ($x_0 = y_0 = z_0 = 0$) is described in $S$ by $y = z = 0$ and by

$$x = -\frac{\beta c}{1 - \beta^2} t \, . \tag{27}$$

This origin is thus seen to move with speed $\beta c/\left(1-\beta^2\right)$, which can exceed $c$, but cannot be superluminal. In fact a light pulse seen from $S$ to propagate in the same direction as $S_0$ has $\theta = \pi$, and thus [using (25)] has speed $c(\pi) = c/(1-\beta)$, which can easily be checked to satisfy

$$\frac{c}{1-\beta} \geq \frac{c\beta}{1-\beta^2}.$$

It is clear from (27) that the velocity of $S_0$ relative to $S$ is not equal and opposite to that of $S$ relative to $S_0$. Such reciprocity holds in the TSR, where it is a consequence of the particular synchronization used, but cannot be expected to hold more generally.

Consider now a third inertial system $S'$ moving with velocity $\beta' c$ and its transformation from $S_0$, which, of course, is given by Eq. (24) with $\beta'$ replacing $\beta$. By eliminating the $S_0$ variables, one can obtain the transformation between the two moving systems $S$ and $S'$:

$$\begin{cases} x' = \dfrac{R(\beta)}{R(\beta')}\left[x - \dfrac{\beta' - \beta}{R^2(\beta)}ct\right] \\ y' = y \qquad ; \qquad z' = z . \\ t' = \dfrac{R(\beta')}{R(\beta)}t \end{cases} \tag{28}$$

We have proposed that (24)-(26)-(28) be called "inertial transformations." In its most general form (28), the inertial transformation depends on two absolute velocities ($\upsilon$ and $\upsilon'$). When one of them is zero, either $S$ or $S'$ coincides with the privileged system $S_0$, and (28) becomes either (24) or (26).

By studying the multiplication properties of the inertial transformations it has been possible to show that they do not form a group. There are no problems with the existence of the identical and inverse transformations, and the associative law can also be satisfied, but it is not always possible to write a meaningful product of two inertial transformations, due to the presence of two absolute velocities $\upsilon$ and $\upsilon'$ in the transformation. If $\Omega(\upsilon,\upsilon')$ denotes the transformation (28), it is easy to understand that the product $\Omega(\upsilon,\upsilon')\Omega(\upsilon'',\upsilon''')$ is no inertial transformation if $\upsilon'' \neq \upsilon'$.

One feature characterizing (24)-(26)-(28) is *absolute simultaneity*: two events taking place at different points of $S$ but at the same $t$ are also judged to be simultaneous in $S'$ (and *vice versa*). The existence of absolute simultaneity does not imply that time is absolute: on the contrary, the $\beta$-dependent factor in the transformation of time gives rise to time-dilatation phenomena similar to those of TSR. A clock at rest in $S$ is seen from $S_0$ to run slower, but a clock at rest in $S_0$ is seen from $S$ to run *faster*, so that both observers agree that motion relative to $S_0$ slows down the rate of clocks. This situation differs from the case of TSR because a meaningful comparison of rates implies that a clock $T_0$ at rest in $S_0$ must be compared with clocks at rest at different points of $S$, and the result is therefore dependent on the "convention" adopted for synchronizing the latter clocks.

Absolute length contraction can also be deduced from (24)-(26)-(28): all observers agree that motion relative to $S_0$ leads to contraction. The discrepancy with the TSR is due again to the different "conventions" concerning clock synchronization: the length of a moving rod can only be obtained by marking the *simultaneous* positions of its end points, and therefore depends on the very definition of simultaneity of distant events.

## 6. Conclusions

We have showed that the Lorentz transformations based on light speed invariance do not satisfy the continuity principle. A different theory satisfying it is available and is, moreover, more attractive, because all the paradoxes of the TSR melt away as soon as one adopts the inertial transformations to describe the physical reality of inertial systems: this is so, for example, for the "block universe" paradox discussed in [7].

We made our choice of synchronization by considering rotating platforms. The main result are Eqs. (18) and (20): the ratio

$$\rho \equiv \frac{c_1(\pi)}{c_1(0)}$$

has been calculated along the rim of the platforms and was shown, under very general conditions, to have the value (18), which in general is different from unity. Therefore the speeds of light parallel and anti-parallel to the disk's peripheral velocity are not the same. For TSR, this is a very serious problem, because a set of platforms with growing radius, but all with the same peripheral velocity, locally approaches better and better an inertial frame. To say that the radius becomes very large with constant velocity is the same as saying that the centripetal acceleration goes to zero with constant velocity. The logical situation is shown in Fig. 1 where one can easily see that TSR predicts a discontinuity at zero acceleration, a sudden jump from the accelerated to the "inertial" reference frames. Whereas all experiments are performed in the real physical world [ $\rho=(c+\upsilon)/(c-\upsilon)$ ], our theoretical physics seems to be "out of this world" [ $\rho=1$ ] !

Very probably the above discontinuity is the origin of the synchronization problems encountered by the Global Positioning System and discussed by Kelly [8]: after all, our Earth is also a rotating platform. More generally, it is the theory of the Sagnac effect on the platform that has always resisted a consistent relativistic formulation [9].

It should be stressed that a non-invariant velocity of light is required for all (but one!) inertial systems. In fact, given any such system and a small region of it, it is always possible to conceive of a large and rotating circular platform locally at rest in that region, and the result (20) must then apply. Therefore, the velocity of light is non-isotropic in every inertial reference frame, with the exception of one ( $S_0$ ), where isotropy has been postulated.

Finally we must also conclude that the famous synchronization problem [10] is solved by nature itself, because all conventions but one lead to an unacceptable discontinuity in the physical theory.

## References

[1] F. Selleri, "Theories 'equivalent' to special relativity - I," first paper presented at this conference.

[2] J. Bailey, K. Borer, F. Combley, H. Drumm, F. Krienen, F. Lange, E. Picasso, W. von Ruden, F.J.M. Farley, J.H. Field, P.M. Hattersley and W. Flegel, *Nature*, **268**, 301 (1977).

[3] We reject the idea that the circumference length depends on the direction in which it is measured: G. Rizzi and A. Tartaglia, *Found. Phys*. **28**, 1663 (1998). To measure distances along the rim between points on it, all we have to do is lay unit rods end to end around the rim, so that they reach from one point to the other, and count them. The number of rods cannot depend on the direction in which they are laid!

[4] F. Selleri, *Found. Phys. Lett*. **10**, 73 (1997); On a Physical and Mathematical Discontinuity in Relativity Theory, in: *Open Questions in Relativistic Physics*, F. Selleri, ed., p. 69, Apeiron, Montreal (1998).

[5] A. Einstein, *Ann. Physik* **35**, 898 (1911).

[6] L. D. Landau and E.M. Lifschitz, *The Classical Theory of Fields*, Fourth revised English edition, Butterworth-Heinemann, Oxford (1996).

[7] F. Selleri, *Found. Phys.* **27**, 1527 (1997).

[8] A.G. Kelly, Synchronization of clock-stations and the Sagnac effect, in: *Open Questions in Relativistic Physics*, F. Selleri, ed., pp. 25-38, Apeiron, Montreal (1998).

[9] G. Sagnac, *Compt. Rend.* **157**, 708 (1913); *ibidem* 1410; P. Langevin, *Compt. Rend.* **173**, 831 (1921); **205**, 304 (1937); E. J. Post, *Rev. Mod. Phys.* **39**, 475 (1967).

[10] H. Reichenbach, *The Philosophy of Space and Time*, Dover Publ., New York, (1958); M. Jammer, "Some fundamental problems in the special theory of relativity," in *Open Questions in Relativistic Physics*, G. Toraldo di Francia, ed., Società Italiana di Fisica, Bologna, and North Holland, Amsterdam (1979).

# Physics Has its Principles

Tom Van Flandern[9]

Physicists and mathematicians have fundamentally different approaches to describing reality. The essential difference is that physicists adhere to certain logical principles, any violation of which would amount to a miracle, whereas the equations of mathematics generally are oblivious to physical constraints. This leads to drastically different views of what is, and what is not, possible for cosmology and the reality we live in.

## Introduction

"Something is wrong with science—fundamentally wrong. Theories keep getting stranger and stranger." [Opening words of preface of the author's book, *Dark Matter, Missing Planets and New Comets*] This is certainly true of physics, which has backed itself into apparent contradictions, leading directly to the dominant Copenhagen view that "there is no deep reality to the world around us."

A reasonable person might ask, "What is the wrong turn that physics has taken to arrive at this predicament?" The answer proposed here is that physics has given up its principles. It has too long consorted with mathematicians, who have no such principles. Mathematics obviously has considerable value as a tool for describing the world. However, a strength of physics historically has been the discipline it brings to mathematics by relating directly to nature. Forgetting this has surely been to the detriment of progress in physics.

## The causality principle

Perhaps most basic of all the principles of physics is the causality principle. In its simplest form, it reads: "Every effect has a cause." In more precise language, it reads: "Every effect has an antecedent, proximate cause." Let's examine these components, and see why each is required.

First, why must every effect have a cause? The answer is so basic that it is practically a matter of definition. The "cause" is whatever makes the "ef-

[9] Meta Research / <tvf@mindspring.com>

fect" happen. If something in the universe changes (an effect), having no "cause" to make it happen is the logical equivalent of magic, a miracle, or the supernatural. Even then, we might think of the will of the magician, miracle worker, or supernatural being as the cause. However, we are not referring here to tricks or illusions, but to events that happen without something making them happen. Even the will of a powerful being cannot produce an effect without having the means to do so. The "means" is the cause, and typically involves force or energy in some form. This point will be clearer when we examine the other two parts of the causality principle.

## No time reversal

"Antecedent" means that a cause must exist in time prior to the effect happening. If their order were reversed, we would still refer to the chronologically first as the cause and the second as the effect. This is because if something were able to change the past, it could create logical contradictions. For example, let A cause B, then let B directly or indirectly eliminate A in the past. Then B could never have come into existence because A, now gone, is what caused B; and so on, in an endless loop of contradiction. So logically, all causes must be antecedent to their effects.

[We ignored the possibility of simultaneous cause and effect because that would require change without benefit of the passage of time. But we consider time to be a measure of change in the universe, making change without time a meaningless concept. Of course, nothing prohibits a cause from operating so close to simultaneously that we lack the ability to measure the short interval by which it precedes the effect. For our purposes here, it is important only that the effect must precede the cause, by however miniscule an amount of time.]

It follows that time travel into the past is not possible. Imagine what it would mean for a person to time-travel into the past, as in an H.G. Wells story. As the person appears in a time where he did not previously exist, that instantly violates any hope for conservation of matter or energy in the universe. Not only has more of both just been added to the past (displacing any substance that existed in that place previously), but the universe continues to have this supplemental mass and energy until their progenitors disappear from the present.

Another problem is that time travel must also involve travel through space. For example, the Earth is continuously traveling through space in its orbit around the Sun, in the Sun's orbit around the Galaxy, and in the Galaxy's motion through the local supercluster. If one could suddenly pop into the universe at a past time, how could one expect to find the Earth in space at that time?

Of course, the main reason why time travel is impossible, and not merely technologically difficult, is that it leads to logical contradictions of the type we described above. Sometimes it is claimed in science fiction that time travel must constrain one's freedom of choice, voluntarily or involuntarily, to prevent changes to the future that would cause a logical contradiction. For example, you might be forbidden or prevented from going back and killing your own grandfather.

However, this ignores that your mere appearance in the past has changed the entire universe forever. When you arrive on Earth in the past, you displace or absorb air molecules in some new way, which changes the course of countless numbers of air molecule collisions, which in turn change countless numbers of other similar events. Eventually, some critical event that depended on air molecules being just so—maybe the timing of when a leaf falls, or whether or not something rolls over a cliff, or whether a roll of dice turns up a one instead of a six—will happen differently than in the original time line. That causes the new time line to begin to diverge from the old at an accelerating pace. Each new event generates many other new events that did not happen before. After enough time, everything becomes affected. So it is impossible for time travel over non-trivial time intervals to avoid eventually changing something in a way that leads to a contradiction. Time travel is therefore disallowed by the principles of physics.

## No true "action-at-a-distance"

"Proximate" means "physically in contact with." An effect can have many remote causes, but must have at least one proximate cause. The alternative would be a condition that one thing be able to affect another without the passage of anything between the two. Once again, this would be the logical equivalent of magic, a miracle, or the supernatural. This condition is called

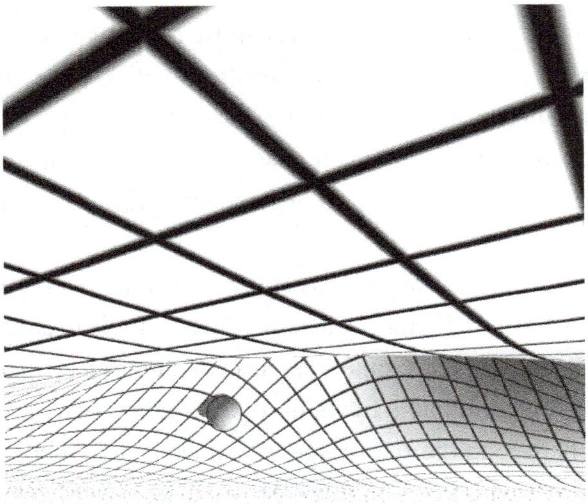

**Figure 1**. Rubber sheet analogy for "curved space-time. Artwork by Starosta.

"action at a distance," and is forbidden by the causality principle because it is logically impossible.

Isaac Newton, whose Universal Law of Gravitation is implicitly based on action-at-a-distance, left no doubt that he considered this a pragmatic approximation of reality when he said: "That one body may act upon another at a distance through a vacuum, without the mediation of any thing else, by and through which their action and force may be conveyed from one to the other, is to me so great an absurdity, that I believe no man who has in philosophical matters a competent faculty of thinking, can ever fall into it." So reality requires that any action be conveyed from a remote cause to a target by means of some sort of action-carriers. It does not require that the carriers be visible or even detectable. But exist they must, and they, or some surrogate carriers, must come into contact with the target to transmit the action. [Those familiar with the extended Zeno's Paradox for matter might object that true contact is impossible when matter is infinitely divisible. However, it suffices that "contact" be the finite limit of an infinite series of increasingly close approaches as one goes ever deeper toward the infinitesimal. This is analogous to crossing a street half way, then half the remaining way, then half again, and so on forever. Although an infinite number of half-the-distance steps are needed, the series nonetheless reaches a finite limit (the other side) in a finite time. For a fuller discussion, see *Dark Matter*, …, Chapter 1.]

Modern physics has introduced the concept of "fields," such as charge around a particle or gravitation around a mass. When the particle or mass moves, its entire field moves with it. However, this cannot happen acausally. For example, the mass may cause adjacent parts of its field to move, which in turn move more distant parts, and so on. This is what happens in any rigid body when one part of it is pushed: a pressure wave propagates through it, conveying the push to all parts of the rigid body. Therefore, fields are not a form of action at a distance. The fact that gravitational fields are seen to update faster than light can propagate (Van Flandern, T., "The speed of gravity—What the experiments say," *Phys. Lett. A* 250, 1-11, 1998) is an argument for faster-than-light propagation of forces, not an argument for action at a distance.

Another modern physics concept is "curved space-time." If such a thing exists and can cause a body to move, then it must itself consist of something tangible or "solid"; *i.e.*, able to act on a body. If so, then it simply constitutes another action carrier updated by other carriers back to the source of gravity. It is reasonable to admit that we know nothing about what constitutes "space-time" or how it carries actions. It is not reasonable to maintain that "space-time" needs no tangible connections to either the source or the target of gravity. Obviously, many mathematical physicists in the field today do not think about "space-time" as tangible in that way. This can lead to some frustrating conversations between people with incompatible perspectives about reality.

To be specific, consider a marble at rest in a curved space-time, as in Figure 1. If at rest, it must remain at rest unless some force acts on it. We are told to visualize that the marble will tend to roll downhill, and this is how "curved space-time" produces the effect we call gravity. However, from a causality perspective, if the rubber sheet or "curved space-time" were located in space without gravity already present under the sheet, the marble would just stay in place on the side of the hill. The existence of curvature, even when time is involved in the curvature, is not a cause of motion. Only a "force" (a conveyor of momentum) can induce new motion. The force is the proximate cause.

## No "creation *ex nihilo*"

"No creation *ex nihilo*" is the principle that something cannot come into existence out of nothing. In a sense, it is another manifestation of the causality principle because such creation would represent an effect without a cause. However, this is a particular case worth considering on its own merits because our primary cosmology today, the Big Bang, begins with the ultimate creation-from-nothing scenario—the mass, space, and time of the entire universe from nothing—as its first step.

Creation *ex nihilo* is forbidden in physics because it requires a miracle. Everything that exists comes from something that existed before, that has grown, or fragmented, or changed form. Growth requires accretion, nourishment, or energy input. Fragmentation ranges from chipping to evaporation to explosion into bits so tiny that we can no longer see or detect them. Changing form includes changes of state, such as solids, liquids, gases, or plasmas.

"Matter" and "energy" may be regarded as simply different forms of the same substance, convertible back and forth. It is easy to visualize matter as exploding into ultra-tiny bits that we might call "energy." But part of that energy consists of the high speeds of bodies. Where does that energy come from? Bodies have small constituents inside atoms that already have high speeds. These constituents may be liberated by an explosion, just as high relative speeds of bodies can be converted into fast constituent motion (heat) during a head-on collision. Even if we could not be specific about how this happened, we could still be certain that energy is not created on the spot from nothing.

So-called spontaneous particle creation from vacuum need not violate this principle because the vacuum is not empty. So called "zero-point energy" is energy of the vacuum, implying that the vacuum is occupied by substance on a scale too small for us to yet detect in any form other than in Casimir-type experiments. The principle only requires that the ingredients from which something is made pre-exist, but not that we can discover them yet.

Religious people might wonder why physics does not admit creation *ex nihilo* as an "act of God," and therefore a valid cause. However, this is a non-economical, and non-testable hypothesis, thereby violating two of the criteria of Scientific Method. Moreover, "acts of God" are a potential explanation for everything, ending the need to investigate further and discover predictable causes. As long as all observations and experiments can be explained without need of miracles—something that has so far remained true—this principle must remain an inviolate guideline. Even if an apparent exception arose, it is difficult to imagine circumstances where a more economical, and therefore more scientific, hypothesis than an act of a Supreme Being would not exist. See also the later section of this paper about "repealing physical principles."

## No "demise *ad nihil*"

The counterpart of not allowing the creation of something from nothing is "No Demise *ad nihil*"; *i.e.*, something cannot become nothing. However finely a thing may dissolve, however undetectable the bits of "energy" into which a thing may explode, if all the individual bits were brought together again with the same ordering, the original thing would be recovered. In other words, nothing has ceased to exist; it has merely changed its appearance or form.

It is conjectured in general relativity (GR) that "black holes" might exist, in which case anything inside an event horizon would be out of communication with the rest of the universe. Such a condition might appear to be the practical equivalent of passing out of existence. However, even for black holes, indications of existence can still be found outside the event horizon in the form of a gravitational field, so the object does continue its influence on the universe.

Nonetheless, as we will shortly consider, objects such as the "black holes" presently attributed to GR are forbidden to exist by the principles of physics (such as the next principle below). A type of astrophysical object for which escape velocity exceeds the speed of light might exist, and we might choose to call that a "black hole." However, such an object would presumably remain in two-way communications with the rest of the universe through the action of faster-than-light particles, and eventually disperse in some way as everything eventually does. But it cannot provide an example of demise *ad nihil*.

## The finite cannot become infinite

The last of the often-self-evident principles of physics we will consider here is "the finite cannot become infinite," and of course *vice versa*. That is because, no matter how many finite things we may collect, their total number and total substance remain finite. Likewise, if something is truly infinite (such as the set of all integers), then no matter how we divide it, at least one piece must remain infinite. And no matter how many equal-sized pieces we divide it into, each will still have an infinite number of components.

A singularity is a point where something has become infinite. In astrophysics, it is a point where matter has collapsed to infinite density and infinitesimal volume. Singularities occur routinely in mathematics. But up to now, whenever a singularity occurs in an equation, some constraint always prevents a singularity from arising in nature. For example, Newton's Universal Law of Gravitation, $a = GM/r^2$, where $a$ is acceleration, $GM$ is the product of the gravitational constant and the mass, and $r$ is the distance from the centre of mass, has a singularity at the origin, $r = 0$. The equation requires acceleration to become infinite at the origin. But in reality, no test particle can ever reach the origin at the centre of mass without first entering into the mass itself, which then changes the acceleration formula in a way that limits acceleration.

A classic example of this principle operating in physics is the "ultraviolet catastrophe." It appeared that the energy of re-radiation of absorbed light should become infinite until Planck realized that such energy must occur in discrete packets, called "photons." In similar manner, every other potential infinity in physics has always led instead to new constraints and improved equations lacking accessible singularities.

Physicists have tended toward the soft view that such infinities have never yet arisen, so perhaps they never will. But the principle is really a logical necessity if energy, force, density, and all physical quantities are viewed as consisting of a finite number of discrete physical components, even if at an undetectable level. Then obviously, no finite sum, however large, can become infinite. This guarantees that any equation containing a singularity will not continue to represent nature in the immediate neighbourhood of that singularity, and that some constraint enforcing singularity-avoidance remains to be discovered in connection with that equation.

Of course, mathematicians are unaccustomed to physical principles and are very comfortable in dealing with singularities in their equations. The mathematicians who have taken over the province of general relativity have therefore, not surprisingly, advocated the existence of real singularities in nature at the centres of "black holes." Einstein himself, as a good physicist, never accepted the concept of black holes, and held that some new constraint would modify his equations in the future. His own words in *Annals of Mathematics*, vol. 40, #4, pp. 922-936 (October 1939, written late in his career while he was at Princeton) are illuminating, showing as they do a respect for physical principles over purely mathematical reasoning:

> *If one considers Schwarzschild's solution of the static gravitational field of spherical symmetry ..., [$g_{44}$] vanishes for $r = m/2$. This means that a clock kept at this place would go at rate zero. Further it is easy to show that both light rays and material particles take an infinitely long time (measured in 'coordinate time') in order to reach the point $r = m/2$ when originating from a point $r > m/2$. In this sense the sphere $r = m/2$ constitutes a place where the field is singular.*
>
> *There arises the question whether it is possible to build up a field containing such singularities with the help of actual gravitating masses, or whether such regions with vanishing $g_{44}$ do not exist in cases which have physical reality... [brief discussion of uncompressible liquids omitted]*
>
> *One is thus led to ask whether matter cannot be introduced in such a way that questionable assumptions are excluded from the very beginning. In fact this can be done by choosing, as the field-producing mass, a great number of small gravitating particles which move freely under the influence of the field produced by all of them together. This is a system resembling a spherical star cluster. ... The result of the following consideration will be that it is impossible to make $g_{44}$ zero anywhere, and that the total gravitating mass which may be produced by distributing particles within a given radius, always remains below*

> *a certain bound. [core of analysis omitted; skipping to conclusions]*
>
> *The essential result of this investigation is a clear understanding as to why the 'Schwarzschild singularities' do not exist in physical reality. ... The 'Schwarzschild singularity' does not appear for the reason that matter cannot be concentrated arbitrarily. And this is due to the fact that otherwise the constituting particles would reach the velocity of light.*
>
> *This investigation arose out of discussions [with Robertson and Bargmann] on the mathematical and physical significance of the Schwarzschild singularity. The problem quite naturally leads to the question, answered by this paper in the negative, as to whether physical models are capable of exhibiting such a singularity.*

Einstein wasn't arguing that the Schwarzschild singularity doesn't exist in the equations, but that it doesn't exist in physical reality. Much as for the case of "the ultraviolet catastrophe," he reasoned that the equations will be shown to be incomplete as observations or experiments approach that limit.

## Corollaries of principles

Many matters of considerable importance follow immediately from the principles of physics. For example, nature has no singularities. If it did, matter could disappear from the universe, violating the no demise *ad nihil* principle while also violating the finite cannot become infinite. The continued action of an external gravitational field after the cause of that field has permanently ceased to communicate with the outside universe is a cause without an effect. And the strange temporal properties of black holes have led to the proposal of "worm holes," which violate the no time reversal principle. Black holes and worm holes are fun science fiction concepts, and are much touted and discussed by mathematical relativists. But no physicist who understands the logical necessity of the principles of physics as descriptors of reality can take such concepts literally.

It follows from these principles that there are no black holes in the traditional relativity sense of event horizons centred on a singularity. This does not preclude highly collapsed states of matter generating a high redshift for light, or possibly no light escape at all. But such objects would continue to have normal gravitational and electrostatic forces and be in two-way communication with the rest of the universe. Some of the fantastic properties of black holes will therefore turn out to be fantasies after all.

Perhaps even more importantly, the physical principles immediately imply that there was no Big Bang at the origin of the universe. The "Big Bang" also violates several physical principles: an effect with no antecedent, proximate cause; no singularities in nature; and no creation *ex nihilo*. If the universe really is expanding—an assumption very much in doubt [see *Dark Matter, Missing Planets and New Comets*, 1999 edition, Chapter 22; reprinted from MRB 3, 25-35 (1994)]—then something must limit how far back that expansion can be projected.

Of course, religions have long taught that the creation of the universe is at least the one major exception to no creation *ex nihilo*. This approach suffers from the difficulties mentioned earlier in connection with ascribing causes to acts of God. As long as it remains clear that viable explanations do exist that require no "acts of God" [see for example *Dark Matter*, …, Chapters 1-2], science will always prefer these because they make reality testable and ultimately predictable, at least to the limits of our understanding.

## Definitions of dimensions

While not a physical principle, the matter of defining dimensions touches on some similar issues in the arena of the mathematicians' approaches versus that of physicists. Mathematicians, lacking physical constraints, are free to imagine or invent unlimited numbers of dimensions, and to describe any properties to them they wish. So one hears often of parallel dimensions, hyper-dimensions, multiple time dimensions, more than three space dimensions, *etc*. It is easy to forget that such ideas are fictional concepts. We have not a single observation or experiment that cannot be fully and completely explained with three dimensions of space, one of time, and one of mass or scale. And despite having many theories of extra dimensions, we have no theoretical requirement for any but the five that are part of our everyday reality. So it is easy to forget that Occam's Razor then requires that we not invent extra physical dimensions unless and until some necessity arises—not convenience, but necessity. Extra mathematical dimensions are fine if they serve a purpose, but should not be confused with physical reality.

A second point about dimensions is that they are scales for the measurement of intervals. As such, they are ordinarily defined to be smooth and linear. Why complicate dimensions unless doing so serves a useful purpose? Moreover, scales for measurement are insubstantial; *i.e.*, they have no substance. Therefore, a dimension cannot be affected by matter or by a force. Consider a common example, often seen in general relativity texts: "curved space." Think of a light ray following that curvature and bending as it passes the Sun's mass. GR suggests we think of the ray path as straight and space as curved. But it would be simpler, as in classical physics, to think of the ray path as curved and the space as straight. In fact, wherever we are in the universe, we can always construct three mutually perpendicular lines, extend them to infinity in both directions, and have all observers in the universe agree that these lines are straight, uniform, and parallel to the straight lines of all other observers, even if they pass near or through large masses. There is clearly no necessity for having curved space, whatever masses or forces may do to light, the vacuum, or other matter. For example, any two points along the curved path of a light ray past a mass can be joined by a taut string, which (if it is strong enough to resist the pull of gravitation and other forces) describes a straight line through space, and a shorter path in space than the ray takes.

Similar remarks apply to time. Clocks may change rates, and they apparently slow down when in a gravitational field or moving relative to such a field. However, the dimension of time can remain as smooth and linear as we please. In much of the 20th century, it was thought that time could not be measured apart from the behaviour of clocks. However, experience with the Global Positioning System (GPS) has shown that, even when clocks move with different relative speeds in different gravitational potentials, all can be synchronized in epoch and rate to hypothetical underlying non-moving clocks in a strictly inertial frame with the gravitational potential projected to any standard height. Then all such clocks will remain permanently synchronized, and make excellent measures of a form of "universal time," compatible with other clocks throughout the universe.

However, if time is not a physical thing that slows down with speed and stops for things moving at the speed of light (as is true in Lorentzian Relativity, but not in Special Relativity), then it follows that the speed of light is not a speed limit for the universe. A hypothetical spaceship traveling at the speed of light might see its atomic clocks stop at that speed, or perhaps even reverse if the spaceship moved faster yet. But time would march forward for the spaceship and the entire universe at the same rate as ever. If the spaceship used chemical propulsion, it might have as much difficulty propelling itself faster than light as a propeller plane would have trouble exceeding the speed of sound. But nothing prohibits this happening in principle if new methods of propulsion, such as gravity, not limited by the speed of electromagnetic radiation, were employed for the purpose.

## Repealing physical principles

It is fun to think of other dimensions, time travel into the past, magic, and numerous other mathematical and/or science fiction concepts. However, it is useful to make a distinction between concepts that are possible, although we are not yet technologically advanced enough to make them happen; *versus* concepts that are now and always certain to be impossible because they lead to logical contradictions. This is reminiscent of the old argument: Can God, who is omnipotent, invent a square circle? The normally accepted answer is that even omnipotence does not enable a Being to devise a contradiction in terms.

In considering this difference, we should acknowledge Clarke's First Law: "Any sufficiently advanced civilization is indistinguishable from magic." The wording of this law notwithstanding, we can tell the difference between advanced technological feats and logically impossible feats. For example, we would not be too startled by an advanced species that had perfected Star-Trek-like teleporters, although that possibility is far beyond what our technology is capable of doing. By contrast, we could rest assured that no species, however advanced, can alter the past. Time travel into the past is a logical impossibility.

Now suppose that we encountered an advanced species that did have the capability to alter the past or violate other physical principles. Ironically, this is not a logical impossibility. For example, we have seen Star-Trek-like holodecks create virtual realities that are essentially indistinguishable from our own reality. Clearly, the programmer can alter the virtual reality program to appear to defy physical principles. Nonetheless, the result is little different from watching a movie about time travel or black holes, even though we might have no awareness that what we sensed was fictional. So if we saw physical principles being violated, we could conclude with some certitude that we were experiencing a virtual reality.

This raises an interesting philosophical challenge: How do we know that our present reality is not a virtual one? The short answer is that, if it is programmed to be faithful to all principles of physics and in other ways realistic, we might well lack any means of being able to tell which type of reality we inhabit. But ultimately, we are forced to act pragmatically and behave as if this reality is non-virtual because the consequences of doing otherwise are painful and catastrophic, to the best of our ability to predict them. [See *Dark Matter*, …, Chapter 20, for a fuller discussion of "truth and reality."] The discovery of a single, clear violation of a principle of physics would change that conclusion. So we can see that a great deal is at stake in adhering to the principles for as long as that remains possible.

## Conclusions

The principles of physics are inviolate rules because any contradiction would be tantamount to magic, a miracle, or the supernatural. The following principles were discussed here:

- Every effect has an antecedent, proximate cause
- No time reversal
- No true action at a distance
- No creation *ex nihilo*
- No demise *ad nihil*
- The finite cannot become infinite

### These corollaries flow from application of the principles:

- Nature has no singularities
- There are no black holes
- There was no Big Bang
- 2-way time travel is impossible

### These corollaries follow from classical definitions of dimensions:

- Extra dimensions are not needed to describe physical reality

- The five ordinary dimensions are always uniform, linear, and universal
- The speed of light is not a universal speed limit
- Discovering a definite violation of a physical principle would bring into question the nature of the reality we inhabit.

# International Atomic Time and the One-Way Speed of Light

Romano Manaresi[10]

We show that the accuracy of the international atomic time (TAI) system imposes no conditions on the one-way speed of light. The TAI is given by a network of atomic clocks distributed around the world that communicate with one another using radio synchronization signals. The synchronization signals sent by a transmitting station always arrive at the receiving station 'on time,' at any time of day and in any season, despite the motion of the earth. For certain authors this means that these signals propagate isotropically (with one-way velocity *c*), even on earth. In fact this may not be so; we shall show that the proper working of the TAI network says nothing about the one-way velocity *c*, as it is consistent with another theory, empirically equivalent to special relativity, in which the one-way speed of light has a directional dependence in moving frames.

## 1. Introduction

There is a whole network of atomic clocks around the world and they are continuously connected *via* radio by synchronization signals: they supply the international atomic time (TAI, Temps Atomique Internationale).

In Sexl and Schmidt's opinion [1] the proper functioning of this system demonstrates that light has the same speed *c* in every direction. They consider two stations with atomic clocks, separated by a distance *d* (measured on the Earth). The first one transmits synchronization signals at regular time intervals to the other one; 12 hours after the first synchronization, due to the Earth's rotation, the radio signal goes in the opposite direction to the previous one, and if its velocity were not constant, a phase-difference between the clocks would be detected. This does not happen, so, they say, the velocity of light is also isotropic on the Earth, which moves through the ether at least at its orbital velocity $v \cong 30$ Km/sec.

The situation, in two generic moments of synchronization, might be the same as shown in Fig.1 (we neglect the Earth's curvature in the path of

[10] v.Tevere 5, I-48017 Conselice RA – Italy. E-mail: Rmanaresi@racine.ra.it

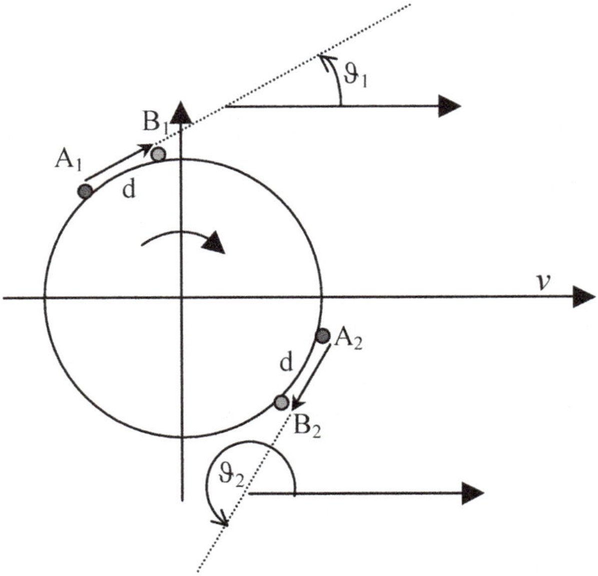


**Fig.1**: Generic situation during two subsequent moments of synchronization between the stations

length *d*). The synchronization of the atomic clocks, experimentally detected, is an objective fact and must be predicted by observers at rest in every system of reference. We will call S the inertial system which moves at velocity *v* with respect to the Earth but, of course, does not rotate.

In this system we will consider the experiment, at first using the Special Relativity (SR) and the Lorentz transformations, then with the theory proposed by F. Selleri with its "inertial transformations" [2].

In such a theory the concepts of absolute space and time are retained and, consequently, the one way velocity of light *does not* have the same value in all inertial systems. Similar conclusions have been obtained by other authors [3]-[6].

## 2. Standard relativistic approach

Light has the same velocity ($c$) in every direction and for everyone, as well as for the two stations, no matter where they are placed.

The time the signal takes will be the same in the two cases of Fig. 1. The movement of the two stations due to rotation, no matter what effect it might have on the clocks, will be the same for both of them, since there are no privileged directions and the relative motions are symmetric. This means that no time difference will be detected by the two clocks.

## 3. Review of the inertial transformations

If we postulate

- (*i*) linearity of the transformations among inertial systems in relative motion,
- (*ii*) homogeneity of space and time,
- (*iii*) existence of an isotropic inertial system $S_0$ (ether) in which the one-way speed of light is $c$ in every direction,
- (*iv*) invariance of the speed of light on closed paths in every inertial system,
- (*v*) time dilation of all the clocks in motion with respect to $S_0$ by the factor

$$\gamma = \frac{1}{\sqrt{1-\beta^2}} \text{ where } \beta = v/c,$$

and $v$ is the clock velocity measured in $S_0$, one obtains the following general transformations between any system S, in "absolute" motion with velocity $v$ (in the direction of $+x$) and $S_0$

$$\begin{cases} x = \gamma\left(x_0 - vt_0\right) \\ t = \dfrac{t_0}{\gamma} + e_1\left(x_0 - vt_0\right) \end{cases}, \tag{1}$$

where $e_1$ is a parameter that remains free. The transformations (1) explain all experimental evidence independently of the value of $e_1$ chosen.

If we presume that light propagates isotropically in one-way mode in every inertial system, we will adopt Einstein's synchronization where $e_1$ will be set as $e_1 = -\beta\gamma/c$. In this case (and only in this one), the principle of relativity is valid in its strong form and the equations (1) become the Lorentz transformations.

There are good reasons to set $e_1 = 0$, and so equations (1) become the inertial transformations [2]

$$\begin{cases} x = \gamma(x_0 - vt_0) \\ t = \dfrac{t_0}{\gamma} \end{cases} \tag{2}$$

This choice, without comparison to any experimental data, gives a different vision of the world, vastly different from the relativistic view.

An absolute space (ether) and an absolute time exist, defined by $S_0$. Rods in absolute motion at velocity $v$ contract by a factor $1/\gamma$ and clocks in absolute motion decrease their rates by the same factor $1/\gamma$, as in SR

But, unlike SR, these effects are not "due to perspective," they are real and absolute: the system in absolute motion measures the *extension* of the rods and the *increase* of the rate of clocks in the ($S_0$) system in absolute rest. These variations are of the same magnitude as the contractions measured by $S_0$. To prove this it is sufficient to invert the equations (2).

In fact, the equations (2) between two $S_1$ and $S_2$ systems, both in absolute motion (with $v_1$ and $v_2$ velocity) become

$$\begin{cases} x_2 = \dfrac{\gamma_2}{\gamma_1} x_1 + \gamma_1\gamma_2(v_1 - v_2)t_1 \\ t_2 = \dfrac{\gamma_1}{\gamma_2} t_1 \end{cases} \quad \begin{cases} x_1 = \dfrac{\gamma_1}{\gamma_2} x_2 + \gamma_1\gamma_2(v_2 - v_1)t_2 \\ t_1 = \dfrac{\gamma_2}{\gamma_1} t_2 \end{cases}, \tag{3}$$

where

$$\gamma_1 = \frac{1}{\sqrt{1 - v_1^2/c^2}} \; ; \; \gamma_2 = \frac{1}{\sqrt{1 - v_2^2/c^2}} . \tag{4}$$

From (3) we find that lengths and durations are transformed by being multiplying by the *ratio between the two system contraction factors*.

Absolute velocities are composed in a Galilean way, but the measurements of velocity in absolute motion systems are rendered false because of

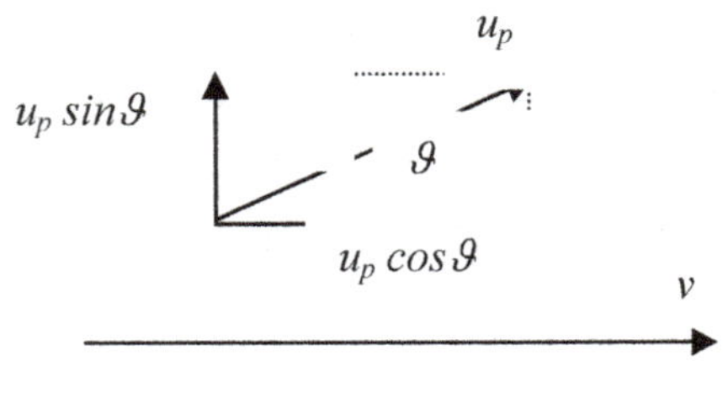


**Fig.2**: Components of a generic velocity

the contraction of rods in the motion direction ($1/\gamma$) and of the slowing down of clocks ($1/\gamma$). Therefore, velocities which are parallel to the absolute $v$ of the moving system S will be measured, in S itself, as increased by a factor $\gamma^2$ (contracted rods and dilated time), while velocities which are perpendicular to $v$ will be increased by a factor $\gamma$ (only dilated time).

A velocity in S in an arbitrary direction can be decomposed into three orthogonal components (Fig.2); we transform it to $S_0$ (absolute velocities) by dividing the $v$-parallel component by $\gamma^2$ and the perpendicular components by $\gamma$, and then sum it (in a Galilean mode) to the velocity $v$ of the S system and compose it again using Pythagoras's theorem.

If a body $p$ moves at a $u_p$ velocity, with respect to the S system in absolute motion at velocity $v$, its absolute velocity $v_p$ will be

$$v_p = \sqrt{\left(v + \frac{u_p \cos\vartheta}{\gamma^2}\right)^2 + \left(\frac{u_p \sin\vartheta}{\gamma}\right)^2}, \tag{5}$$

where $\vartheta$ is the angle between $u_p$ and $v$.

As a consequence of equation (5), the velocity of light is isotropic in $S_0$ only: for the system S in absolute motion, replacing $u_p$ and $v_p$ by $c_{(\vartheta)}$ and $c$ respectively, one gets

$$c_{(\vartheta)} = \frac{c}{1+\beta\cos\vartheta}. \tag{6}$$

Of course, from (6) we can obtain a velocity on closed paths equal to $c$ in every inertial frame.

## 4. Application to the rotating Earth

The two synchronization signals of Fig.1 propagate, in S, respectively, at

$$c_1 = \frac{c}{1+\beta\cos\vartheta_1} \text{ and } c_2 = \frac{c}{1+\beta\cos\vartheta_2}, \tag{7}$$

while the arrival station moves away at the rotation velocity.

But S (which synchronized its clocks according to the inertial transformations) does not detect a uniform rotation velocity. Let us explain why.

We set in S a segment P-Q with a unitary length, and when the terrestrial station A passes (at velocity $u_A$) close to P, P itself sends out a light signal (at velocity $c_A$). In Q we measure the time lag between the arrival of the light signal and the passage, still in Q, of station A.

This delay will be the difference between the propagation times: $1/u_A - 1/c_A$ . This measurement is made with one clock only, so the result cannot depend on the synchronization used, and must be the same as would be obtained according to SR. Therefore, this delay does not depend on the position or the direction of the velocities used in the measurements.

So, $u_1$ and $u_2$ being the two velocities of Earth on two generic positions 1 and 2, the time differences must be equal,

$$\frac{1}{u_1} - \frac{1}{c_1} = \frac{1}{u_2} - \frac{1}{c_2}. \tag{8}$$

Because $c_1$ is different from $c_2$, $u_1$ will also be different from $u_2$, as is immediately clear from (8). Substituting (7) in (8), one easily gets

$$\frac{c - \beta u_1 \cos\vartheta_1}{u_1} = \frac{c - \beta u_2 \cos\vartheta_2}{u_2}, \tag{9}$$

a result which will soon be useful.

Let us apply what we have found to a physical situation.

According to (5), if $u_1$ and $u_2$ are the rotation velocities of the stations (measured in S) in the two cases, the corresponding absolute velocities $v_1$ and $v_2$ will be

$$\begin{aligned} v_1 &= \sqrt{\left(v + \frac{u_1 \cos\vartheta_1}{\gamma^2}\right)^2 + \left(\frac{u_1 \operatorname{sen}\vartheta_1}{\gamma}\right)^2} \\ v_2 &= \sqrt{\left(v + \frac{u_2 \cos\vartheta_2}{\gamma^2}\right)^2 + \left(\frac{u_2 \operatorname{sen}\vartheta_2}{\gamma}\right)^2} \end{aligned}. \tag{10}$$

A short direct calculation of $\gamma_1^{-2}$ and $\gamma_2^{-2}$ starting from (4) and (10) gives

$$\begin{cases} \gamma_1^{-2} = \dfrac{1-\beta^2}{c^2}\left[\left(c - \beta u_1 \cos\vartheta_1\right)^2 - u_1^2\right] \\ \gamma_2^{-2} = \dfrac{1-\beta^2}{c^2}\left[\left(c - \beta u_2 \cos\vartheta_2\right)^2 - u_2^2\right] \end{cases}, \tag{11}$$

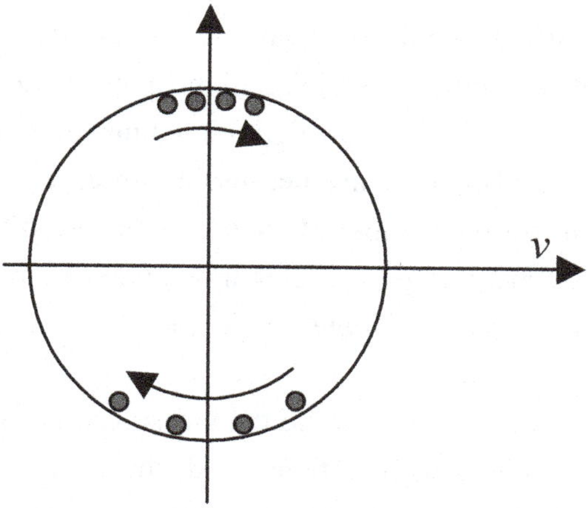


**Fig.3**: Inhomogeneity of a rotating and translating body

whence, taking (9) into account

$$\gamma_1^{-2} = \frac{u_1^2}{u_2^2}\gamma_2^{-2}. \qquad (12)$$

Therefore

$$\gamma_1 u_1 = \gamma_2 u_2 . \qquad (13)$$

The condition (13) allows us to reason as follows.

The contractions of bodies in motion are due to the absolute velocities, and these velocities, in the cases considered in Fig.1, are different: in the first case rotation contributes to an increase of $v_1$, while in the second case it contributes to a decrease of $v_2$ (in the whole upper hemisphere of the picture the absolute velocities are larger than in the lower hemisphere).

So the contractions are different and, as they are real as well, they produce an inhomogeneity of the body itself (Fig. 3): the distances among the atoms are smaller in position 1 (and in the whole upper hemisphere) than in position 2 (in the whole lower hemisphere). What has to be kept uniform is not the velocity of rotation but the "flow," the flux of matter, the number of atoms that traverse any given section in the unit of time. Equation (13) ensures exactly this.

The quantity of matter traversing an arbitrary section (for instance the $v$ perpendicular axis of Fig.3, which passes through the centre) in the unit of time has to be constant. In the opposite case, the rotation would have a translation component, which would shift and accumulate (either backward or forward) some matter in time, thus violating the conservation of the quantity of motion of the Earth in S (which, by definition, must be zero).

This condition is also necessary because an observer on S, who counts the number of atoms (or the number of the unit rods placed longitudinally on the Earth's circumference) he sees passing in the unit of time, finds the same value in each point: *when the atoms (rods) are more contracted, they go slower.*

It is important to notice that, while the measurement of the rotation velocity $u$ depends on the transformations used, the isotropy of the flow of atoms (or rods) around the circumference is *objective* and cannot depend on the transformations used; hence the result must be the same as would be obtained by a "relativistic" observer who, of course, measures an isotropic flow.

## 5. The time differences

Lastly, let us apply the inertial transformations (3) in S and calculate the times taken by the two synchronization signals to travel the distance $d$ as ratios between distances between the two stations and the velocities of light:

$$t_{c1} = \frac{\gamma}{\gamma_1}\frac{d}{c_1 - u_1}; \; t_{c2} = \frac{\gamma}{\gamma_2}\frac{d}{c_2 - u_2}. \tag{14}$$

These times will be registered by the stations on the Earth, the former slowed down by $\gamma/\gamma_1$ and the latter by $\gamma/\gamma_2$ . Their difference will be (clock desynchronization)

$$\Delta t'_c = \frac{\gamma}{\gamma_1}t_{c1} - \frac{\gamma}{\gamma_2}t_{c2} = \frac{\gamma^2}{\gamma_1^2}\frac{d}{c_1 - u_1} - \frac{\gamma^2}{\gamma_2^2}\frac{d}{c_2 - u_2}, \tag{15}$$

as station A and the second synchronization signal are in *advance* by this value.

But there is a *second phenomenon which also produces a desynchronization*, and to quantify that we will again place ourselves in S.

The paths followed by the stations, due to the rotation of Earth, are not the same: the symmetry is broken by the presence of a preferred direction, that of the absolute $v$ of translation with which the rotation velocity composes. The different paths followed by the stations produce different rates of

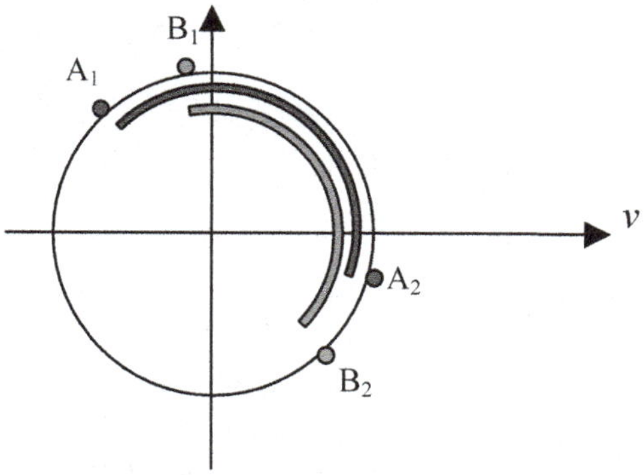


**Fig.4**: The absolute paths of the two stations are different

slowing of their own clocks, the absolute velocities being different. This point is crucial (Fig. 4)

The segment $B_1$-$A_2$ is common to the two paths and does not introduce any difference.

The segment $A_1$-$B_1$ is travelled only by station A at a faster (absolute) velocity, compared to the segment $A_2$-$B_2$ which is travelled only by the station B at a lower (absolute) velocity. The times marked by the two clocks, in these two segments, will be different: station A, which travels faster, will show a *delay*.

The observer in S calculates the times taken to cover the two segments, still as ratios between distances and velocities, thus

$$t_{u1} = \frac{\gamma}{\gamma_1}\frac{d}{u_1}\,;\; t_{u2} = \frac{\gamma}{\gamma_2}\frac{d}{u_2}\,. \tag{16}$$

As we did earlier, we calculate the difference between these times, but now considered in the terrestrial stations,

$$\Delta t'_u = \frac{\gamma}{\gamma_1}t_{u1} - \frac{\gamma}{\gamma_2}t_{u2} = \frac{\gamma^2}{\gamma_1^2}\frac{d}{u_1} - \frac{\gamma^2}{\gamma_2^2}\frac{d}{u_2}\,. \tag{17}$$

Both the distances between atoms (length of rods) and the rates of clocks in a system that rotates and translates relative to the ether are subject to cyclic variations due to the composition of the rotation and translation movements.

We now show that the two time delays cancel, so that

$$\Delta t'_c + \Delta t'_u = 0\,. \tag{18}$$

From the definitions (15) and (17) one easily gets

$$\Delta t_c' + \Delta t_u' = \frac{d\gamma^2}{\gamma_1^2}\left[\frac{1}{c_1 - u_1} + \frac{1}{u_1}\right] - \frac{d\gamma^2}{\gamma_2^2}\left[\frac{1}{c_2 - u_2} + \frac{1}{u_2}\right], \tag{19}$$

which is the same as

$$\Delta t_c' + \Delta t_u' = \frac{d\gamma^2}{\gamma_1^2 u_1^2}\left[\frac{u_1^2}{c_1 - u_1} + u_1\right] - \frac{d\gamma^2}{\gamma_2^2 u_2^2}\left[\frac{u_2^2}{c_2 - u_2} + u_2\right]. \tag{20}$$

The first (second) term in parenthesis in (20) is equal to the inverse left hand side (right hand side) of Eq. (8). Therefore these two terms are equal. The multiplying factors are also equal because of Eq. (13). Thus, the right hand side of (20), being the difference of two equal terms, vanishes and Eq. (18) holds.

Therefore, the two stations will not, in any case, detect any out-of-phase condition between the clocks. The different times taken by the two synchronization signals are compensated by the variations of the rates of the two clocks, which are in an absolute motion composed of the constant translation velocity $v$, plus the rotation velocity $u$ (variable in direction and modulus).

Note: the well-known Sagnac effect has not been taken into account because in the situations considered here, it contributes with *constant* and equal delays at every position of the stations. Since we have considered the *difference* between the delays in different positions, the Sagnac effect is irrelevant here.

## 6. Conclusions

With Lorentz's transformations, $c$ is isotropic in every system and the clocks must be synchronous.

With the inertial transformations, we have two out of phase effects on the clocks: the first because of the $c$ anisotropy, the other because of the variations in the clocks' rates, due to their different absolute $v$. The two effects are equal and opposite and the clocks appear to be synchronized again.

Sexl and Schmidt demonstrate that Galileo's transformations do not function, but Lorentz's transformations are not the only ones that can explain these facts. They are also explained by Selleri's transformations.

The two theories are equivalent. The accurate functioning of the Global Positioning System does not say anything about one-way light-speed and cannot establish which transformation is "true."

## Acknowledgment

I would have many reasons to be grateful to Prof. F. Selleri, but I will mention only one: he made me understand that no matter how fundamental and tested a theory is, it must not become an object of faith.

## References

[1] R. Sexl, H.K. Schmidt, *Raum-Zeit-Relativitat*, Rowohlt Taschenbuch Verlag, Hamburg (1978).

[2] F. Selleri, *Found. Phys.* **26**, 641 (1996); *Found. Phys. Lett.* **9**, 43 (1996).

[3] R. Mansouri and R. Sexl, *General Relat. Gravit.* **8**, 497, 515, 809 (1977)

[4] C.M. Will, *Phys. Rev.* **D 45**, 403 (1992)

[5] F. Goy, *Found. Phys. Lett.* **9**, 165 (1996)

[6] J.R. Croca and F. Selleri, *Nuovo Cim.* **114 B**, 447 (1999).

# New Empirical Clues for the Factor 1.23

A. Rubčić and J. Rubčić[11]
H. Arp[12]

The quantization period of the intrinsic redshift of quasars is characterized by the factor 1.23. The origin of this constant is not yet explained, but it could be of fundamental significance if found in other physical phenomena. Because the radiation emitted from quasars is a consequence of the interaction between elementary particles, the masses of leptons, quarks, mesons and baryons are here first investigated. It is found that their masses are related to each other by integer numbers as $m_k = m_o B^k$ (where $m_o$ is the mass of the first particle in a group and $k$ is the integer number). The quantities $B$ depend on the particular group which the particle belongs to and are found to be simple functions of the fine structure constant, $\alpha$, and the quasar redshift factor, $F \sim 1.23$.

The next larger systems are the atomic nuclei. The correlation of atomic weight $A$ with atomic number $Z$ of elements results in an empirical formula with $F = 1.2375$. The masses of large gravitational systems are also examined, particularly the solar-system in which there are five subsystems: the first one is the Sun with its planets, and the next four are the planets with their systems of satellites. The correlation of the central mass $M_C$ with the sum of the masses $m_S$ of all bodies orbiting the central one, is of the form $m_S = \text{const} \cdot (M_C)^F$, where $F$ is again about 1.22.

The Karlsson formula for the preferred redshifts observed in quasars is derived directly from the Bohr model of the radiating atom using the assumption that electron mass evolves in steps such that $m_{e,k} = m_e/F^k$. Here $F$ is observationally measured to be close to 1.23.

The relationships found here between the fine structure constant and $F$ and the radiation and gravitational properties of physical systems on all scales would appear to be numerically significant. If this is so, it is hoped this will eventually lead to an understanding of the basic cause of quantization on all scales.

[11] Department of Physics, University of Zagreb, Bijenička c. 32, 10001 Zagreb, Croatia (rubcic@sirius.phy.hr)
[12] Max-Planck-Institut für Astrophysik, Karl-Schwarzschild-Strasse 1, 95740 Garching bei München, Germany (arp@mpa-garching.mpg.de)

## Introduction

It has been shown [1] that the factor of intrinsic redshift quantization for the quasars is 1.23. The mathematical formulation is $(1 + z_2) = 1.23(1 + z_1)$, where $z_2$ is the next higher redshift from $z_1$, giving the redshift periodicities of: $z = 0.061$, 0.30, 0.60, 0.91, 1.41, 1.96, *etc*. These values are in a good agreement with observed quasar redshifts. Numerous examples may be found in Arp's book [2]. However, what the factor 1.23 (hereinafter denoted as $F$) represents, or where it comes from is still unexplained. One possibility is that the mass of electrons in quasars is not the standard mass of an electron known from our laboratory measurements, but smaller, increasing in quantized steps as the matter of quasars ages [2,3]. But it would be important to test whether the factor $F$ exists in other physical systems. First, one might expect the appearance of $F$ in elementary particles. If it is found there, the analysis $F$ could be extended to larger systems.

Our analysis is presented in the following order:

- the laws for masses of elementary particles, *i.e.* leptons, quarks, mesons and baryons,
- dependence of the atomic weight $A$ on the atomic number $Z$ of elements,
- masses in solar- and extra-solar planetary systems,
- discussion and conclusion.

Note that the following approach is basically a numerical analysis of observed values with the purpose of suggesting and testing future theoretical interpretations.

## The factor *F* and elementary particles

Table 1 gives a list of masses in units MeV/$c^2$ of elementary particles. The logarithms of these values are plotted in Fig. 1. Quarks $u$, $c$, $t$ and $d$, $s$, $b$ are presented separately due to their different charges $\frac{2}{3}$ and $-\frac{1}{3}$ of elementary charge $e$, respectively. Also the upper and lower limit of mass for a given quark is added. In all there are five groups, and the members in a particular group may be described by almost equidistant values. Consequently, one is able to attribute an integer number to each value of mass within a group. Thus, $u$, $c$ and $t$ quarks have assigned integer numbers $k$ equal to 1, 2 and 3, respectively. For leptons $e$, $\mu$ and $\tau$, the integers are 1, 3, and 4. Neutrinos are

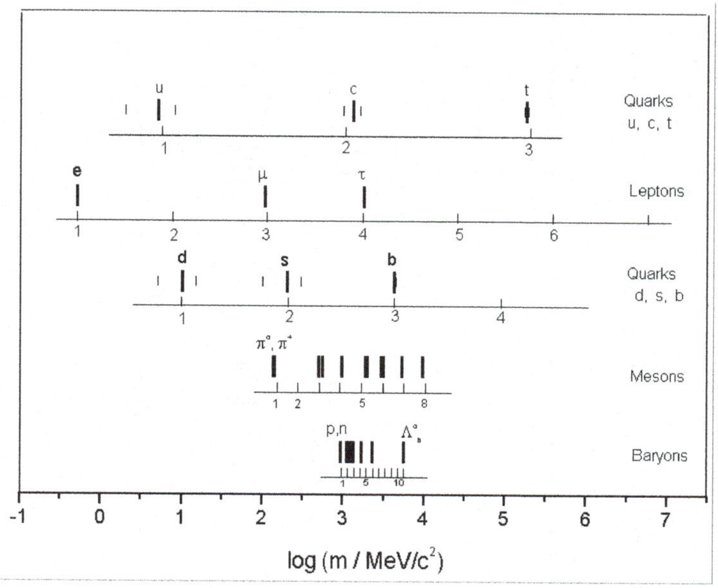


**Figure 1**. Logarithm of the masses of elementary particles. Quarks are separated according to their charges. Small tick marks (denoted by numbers) attached to horizontal lines are the best fit of equation for masses $m_k = m_o B^k$, where $k$ is the integer number.

not shown in Fig. 1 due to the larger scale (log $m$ extends from –5 to 1), but the numbers $k$ are the same as for $e$, $\mu$ and $\tau$.

Such an arbitrary assignment is useful in obtaining the mathematical form for increasing the mass with increasing integer number $k$. The values of $k$ in each group are shown in Fig. 1. Once the numbers $k$ are determined, the linear regression gives the desired equation for increasing mass in a given group. The mean value of the interval (shown in Table 1) for a given quark is used in the calculation. The tick marks on the horizontal axes give the equidistant values calculated by the following equation:

Quarks $u, c, t$ $\quad \log m_k = (-1.53357 \pm 0.17064) + \mathrm{k}(2.27815 \pm 0.07899)$ (1)

Neutrinos $\quad \log m_k = (-7.07854 \pm 0.06236) + k(2.09077 \pm 0.02118)$

Leptons $e, \mu, \tau$ $\quad \log m_k = (-1.47848 \pm 0.04957) + k(1.17818 \pm 0.01684)$ (2)

Quarks $d, s, b$ $\quad \log m_k = (-0.32197 \pm 0.01959) + k(1.31674 \pm 0.00917)$

Mesons $\quad \log m_k = (1.91635 \pm 0.02306) + k(0.26186 \pm 0.00445)$ (3)

Baryons $\quad \log m_k = (2.90512 \pm 0.00694) + k(0.07676 \pm 0.00159)$

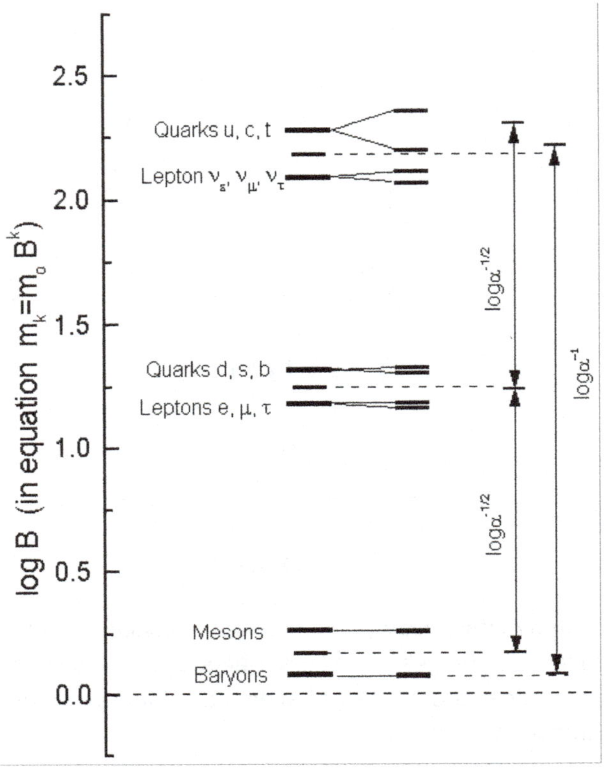


**Figure 2**. The slope of straight lines defined by equation $\log m_k = \log m_o + k \log B$. Mean values of the groups: ($u$,$c$,$t$ and $\nu_e$, $\nu_\mu$, $\nu_\tau$), ($d$,$s$,$b$ and $e$,$\mu$, $\tau$) and (mesons and baryons), are indicated in col.1. Deviations of log $B$, according to Eqs.(1-3) are shown in col.2. Vertical lines at right represent the logarithm of the fine-structure constant $\alpha$.

These equations, written in a logarithmic form $\log m_k = \log m_o + k \log B$, lead to an integer law of elementary masses

$$m_k = m_o B^k, k = 0,1,2,3\ldots \quad (4)$$

In Eqs.(1-3) one notices three groups of the values log $B$.

In Eq.(1) log $B$ is of the order $\log \alpha^{-1} = \log 137.0359895 = 2.1368346$, where $\alpha$ is the fine-structure constant. Similarly, for Eq.(2), $\log \alpha^{-1/2} = 1.06841732$, and for Eq.(3) $\log \alpha^0 = 0$. Values of log $B$ are plotted in Fig. 2, showing the splitting around the mean values for all three groups (col.1). The deviation of log $B$, following Eqs. (1-3) are also shown (col.2). Mean values of log $B$ are presented in the last three lines (col.1) and they are: $2.18446 \pm 0.09369$, $1.24696 \pm 0.06978$ and $0.16932 \pm 0.09255$. Vertical lines

**Table 1. Masses of elementary particles: quarks, leptons, mesons and baryons.**

| Particle | Mass ($MeV/c^2$) | Particle | Mass | Particle | Mass |
|---|---|---|---|---|---|
| | Quarks | | Mesons | | Baryons |
| $u$ | 2 - 8 | $\pi_0$ | 134.9764 | $p^+$ | 938.27231 |
| $c$ | 1000 - 1600 | $p^+$ | 139.56995 | $n^0$ | 939.56563 |
| $t$ | 168000 - 192000 | $\eta$ | 547.45 | $\Lambda$ | 1115.684 |
| | | $\eta'$ | 957.77 | $\Sigma^+$ | 1189.37 |
| $d$ | 5 - 15 | $K^0$ | 497.672 | $\Sigma^0$ | 1192.55 |
| $s$ | 100 - 300 | $K^+$ | 493.677 | $\Sigma^-$ | 1197.436 |
| $b$ | 4100 - 4500 | $D^+$ | 1869.4 | $\Xi^0$ | 1314.9 |
| | Leptons | $D^0$ | 1864.6 | $\Xi^-$ | 1321.32 |
| $e$ | 0.51099907 | $\eta_c$ | 2979.8 | $\Omega^-$ | 1672.45 |
| $\mu$ | 105.658389 | $D_s^+$ | 1968.5 | $\Lambda_c^+$ | 2285.1 |
| $\tau$ | 1777.0 | $B^+$ | 5278.9 | $\Lambda_b^0$ | 5641.0 |
| | | $B^0$ | 5279.2 | | |
| $\nu_e$ | ~ 10 eV | $B_s^+$ | 5369.3 | | |
| $\nu_\mu$ | < 0.17 MeV | $J/\Psi$ | 3096.8 | | |
| $\nu_\tau$ | < 18.2 MeV | $\Upsilon$ | 9460.37 | | |

Note: Data are taken from: a) *Phys.Rev.D-Particles and Fields*, Vol.50. Nr.3, Third Series, (1994), b) C.Caso *et al.*, *European Physical Journal*, C3, 1 (1998).

show $\log \alpha^{-1/2}$ (starting from the mean value of mesons and baryons: 0.16932), or $\log \alpha^{-1}$. Splittings about the mean values define the factor $F$. This assumption gives $\log F = 0.09369$, 0.06978 and 0.09255. It follows that $F = 1.2408$, 1.1743 and 1.2375 with an average value

$$F = 1.218 \pm 0.029$$

The deviation of $F$ is only 2.4%, but it is still too much. The most reliable value is expected to be that obtained from mesons and baryons

$$\log F = 0.09255 \qquad F = 1.2375 \pm 0.0087. \tag{5}$$

In order to illustrate the accuracy of Eqs. (1-3), consider, for example, the leptons $e$, $\mu$, $\tau$ Eq.(2). The resulting masses are 0.501, 113.8 and 1714, respectively. Deviations from the current values are –2.0%, 7.7% and –3.5%, which are too large to be acceptable. This means that Eqs. (1-3) show good general behaviour, but refinement is necessary.

Eq. (4) for two different numbers $k$ gives $m_{k'}/m_k = B^{k'-k}$. It follows that $m_\mu/m_e = B^2$. From the previous statement that for leptons log $B$ is approximately equal to log $\alpha^{-1/2}$ one may put $B = F\alpha^{-1/2}$. It comes out that factor $F$ in the first approximation is

$$F = \left[\frac{m_\mu}{m_e}\alpha\right]^{1/2} = 1.22836\,. \qquad (6)$$

Similarly,

$$F = \left(\frac{m_\tau}{m_e}\alpha^{3/2}\right)^{1/3} = 1.2942 \quad \text{and} \quad F = \left(\frac{m_\tau}{m_\mu}\right)\alpha^{1/2} = 1.4367\,.$$

Dispersion of $F$ is still too large, and a further refinement is necessary. It is easy to see that $F^{1/4} = (1.22836)^{1/4} = 1.05276$, and therefore $1.2942 = 1.22836 \cdot 1.05276 = 1.2932 = F^{5/4}$. Also, 1.4367 is very close to $F^{7/4} = 1.4332$. The second approximation is then

$$\frac{m_\mu}{m_e} = \left(\frac{F}{\sqrt{\alpha}}\right)^2 \quad \frac{m_\tau}{m_e} = \left(\frac{F}{\sqrt{\alpha}}\right)^3 F^{3/4} \quad \frac{m_\tau}{m_\mu} = \left(F\sqrt{\alpha}F^{3/4}\right). \qquad (7)$$

From known masses (Table 1) one obtains by Eq.(7)

$$F = 1.2292 \pm 0.0008. \qquad (8)$$

A still smaller error in $F$ is obtained in the third approximation using reduced masses

$$(m_e + m_\mu)/m_e = F^2/\alpha \qquad F = 1.2313$$

$$(m_e + m_\mu + m_\tau)/(m_e + m_\mu) = F^2/\alpha^{1/2} \qquad F = 1.2309 \qquad (9)$$

$$(m_e + m_\mu + m_\tau)/m_e = F^4/\alpha^{3/2} \qquad F = 1.2311.$$

The mean value for $F$ is

$$F = 1.2311 \pm 0.0002.$$

The comparison of ratios of observed masses and calculated masses is

| | $m_\mu/m_e$ | $m_\tau/m_e$ | $m_\tau/m_\mu$ |
|---|---|---|---|
| Observed | 206.7682609 | 3477.501437 | 16.81835221 |
| Calculated | 206.693 ± 0.067 | 3477.2 ± 2.3 | 16.8230 ± 0.0058 |

The explicit form for the masses $m_\mu$ and $m_\tau$ from Eqs.(9) is

$$m_\mu = m_e\left(\frac{F^2}{\alpha}-1\right) \quad m_\tau = m_\varepsilon\left[\frac{F^2}{\alpha}\left(\frac{F^2}{\alpha^{1/2}}-1\right)\right]. \tag{10}$$

According to Eqs. (4) and (10) the coefficient $B$ is a function of the factor $F$ and the fine-structure constant $\alpha$, *i.e.*,

$$B = f(F,\alpha). \tag{11}$$

It is interesting to note that the lifetime $\tau_{\mu+}$ of the free muon $\mu^+$ and the lifetime $\tau_M$ of a bound muon in muonium $M = (\mu^+ e^-)$ are connected by [4]

$$\tau_M = \tau_{\mu+}\left(1+\frac{1}{2}\frac{\alpha^2 m_e}{m_\mu}\right),$$

which may be written, using $F^2 = (m_\mu/m_e)\alpha$, as

$$\tau_M = \tau_{\mu+}\left[1+\frac{1}{2}\left(\frac{\alpha}{F}\right)^4\right].$$

## Quarks *d*, *s*, *b*

From Table 1 we read

$$m_d = 10 \pm 5(\mathrm{MeV}/c^2)$$

$$m_s = 200 \pm 100$$

$$m_b = 4300 \pm 200.$$

The ratios of masses are $m_s/m_d = 20$, $m_b/m_s = 21.5$ and $m_b/m_d = 430 = 20.74^2$. The mean value of these ratios is 20.747, and following the previous analysis we immediately concludes that the appropriate value for $B(F,\alpha) = F^3/\alpha^{1/2} - 1$. If we suppose that $F = 1.23$, then $B = 20.78375$. The ratio of masses may be then written as

$$\frac{m_s}{m_d} = \frac{F^3}{\alpha^{1/2}} - 1, \quad \frac{m_b}{m_d} = \left(\frac{F^3}{\alpha^{1/2}} - 1\right)^2, \quad \frac{m_b}{m_s} = \frac{F^3}{\alpha^{1/2}} - 1. \tag{12}$$

If we take $m_b = 4300$ and $F = 1.23$, then it follows that $m_d = 10$ and $m_s = 207$. A compact form for quarks *d*,*s*,*b* is

$$m_k = m_b \left( \frac{F^3}{\alpha^{1/2}} - 1 \right)^{k-1}, \quad k = 1,2,3 \quad F = 1.23 . \tag{13}$$

with $k = 1 : m_1 = m_d$, $k = 2 : m_2 = m_s$, and $k = 3 : m_3 = m_b$. It should be noted that for quarks an additional refinement is not necessary due to large uncertainties in the $d$ and $s$ masses. The deviation of masses calculated by Eq.(13) from the mean values is 0%, 3.5% and 0%, or, on average, nearly 1%.

## Quarks *u, c, t*

From Table 1 we read

$$m_u = 5 \pm 3$$

$$m_c = 1300 \pm 300$$

$$m_t = 180000 \pm 12000.$$

Following the same procedure as before we obtain

$$m_k = m_u \left( \frac{F^3}{\alpha} - 1 \right)^{k-1}, \quad k = 1,2,3 \quad F = 1.23 \tag{14}$$

Using $m_t = 180000$ and $F = 1.23$ we obtain $m_u = 6.4$ and $m_c = 1074$, which is within the error of the values given above.

## Mesons

Eq. (3) for mesons after an antilog operation becomes $m_k = 82.48026 \, (1.82751)^k$. Coefficient $B = 1.82751 = (1.22261)^3$, which is very close to $(F - \alpha)^3 = (1.22270)^3 = 1.82793$.

Consequently, we may write

$$m_k = m_1 \left[ (F-1)^3 \right]^{k-1}, \quad k = 1,2,3 \quad F = 1.23 . \tag{15}$$

## Baryons

Eq. (3) for baryons after an antilog operation becomes $m_k = 803.74818 \, (1.19333)^k$.

We e may write this simply as

$$m_k = m_1 (F')^{k-1}, \quad k = 1,2,3 \quad F' = 1.1933 . \tag{16}$$

Lee [5] argues that 1.19 is a characteristic number in the formation of gravitational systems, equally valid as the factor $F = 1.23$. However, coefficient 1.1933 may be written as $(F - (3/7)\alpha^{-1/2}) = 1.1934$ and

$$m_k = m_1\left[F - \frac{3}{7\alpha^{1/2}}\right]^{k-1}, \quad k = 1,2,3 \quad F = 1.23.$$

## Note on fundamental fermions

The masses of all fundamental fermions may be calculated using the mass of the electron. The law for increasing the mass within a group is given by Eq. (4). Eqs. (10) also determine the mass of the muon and $\tau$-lepton from the electron mass, and it is possible to extend these formulae to quarks. Taking the electron mass as the origin, the three straight lines for $(e,\mu,\tau)$, $(e,d,s,b)$, and $(e,u,c,t)$ have a common point: the electron mass $m_e$. It is presented in Fig. 3. If the electron mass is held fixed during fitting, *i.e.*, only the slope of lines is subjected to change, the following equations result:

$$e,u,c,t \qquad m_k = m_e\left(\sqrt{\frac{F}{\alpha}}\right)^k \quad k = 0,1,3,5 \quad F = 1.2257 \tag{17}$$

$$e,d,s,b \qquad m_k = m_e\left(\frac{F^2}{\sqrt{a}}F^{2/3}\right)^k \quad k = 0,1,2,3 \quad F = 1.2253 \tag{18}$$

$$e,\mu,\tau \qquad m_k = m_e\left(\frac{F^2}{\sqrt{a}} + \frac{1}{2}\right)^k \quad k = 0,2,3 \quad F = 1.2309 \tag{19}$$

The mean value of factor $F$ from Eqs. (17-19) is $\langle F\rangle = 1.2273 \pm 0.0024$. Eq. (17-19) are different compared with Eqs.(10,13 and 14) and according to this, one can see how sensitive the results are to the starting assumptions. However, it is important that the dispersion of the slopes of the straight lines for three groups of fundamental fermions in Fig. 3 are limited by the functions $m_k = m_e(1/\alpha^{1/2})^k$ and $m_k = m_e(F^3/\alpha^{1/2})^k$, as shown by the dashed lines. Eq. (17) for $e,u,c,t$ particles includes $k = 0,1,3,5$. In order to have successive numbers $k$, Eq. (17) may be written in another form $m_k = m_e\,[\sqrt{(F/\alpha)}]^{2k-\delta(k)}$, $k = 0,1,2,3$, where $\delta(k) = 0$ for $k < 1$ and 1 for $k \geq 1$.

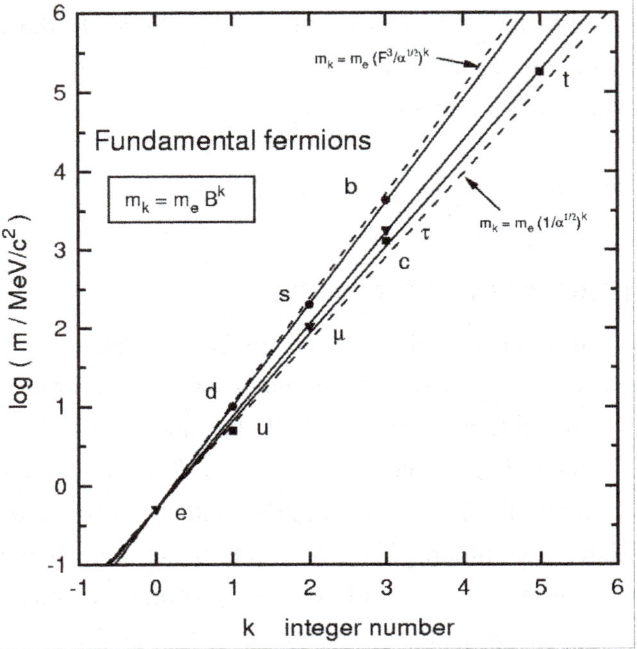


**Figure 3**. Log-lin correlation of the masses of fundamental fermions with the integer number *k*. The electron mass is held fixed, while slopes were subjected to fitting.

Note that all functions $B = f(F,\alpha)$ fall in the interval between $\alpha^{-1/2}$ and $F^3\alpha^{-1/2}$. These functions presently have no theoretical basis. However, they are suggested from experimental data and are very simple. It is hoped that the physical reasons for $B = f(F,\alpha)$ and the forms which the function takes might be understood in the future.

## 2. *A–Z* correlation

In the previous section the elementary particles were considered and an empirical necessity for introduction of the factor $F \sim 1.23$ has been demonstrated. Now, it is a natural extension to examine the atomic nuclei, as the next larger systems. The atomic weight $A$ depends on an integer number, *i.e.*, atomic number $Z$. It is well known that mean atomic weight $A$ depends on $Z$ as a power law. It should be interesting to see this correlation in connection with the factor $F$.

In the physics of nuclei, the semi-empirical atomic mass formula is based on the liquid drop model of the nucleus [6]. In this formula, the term

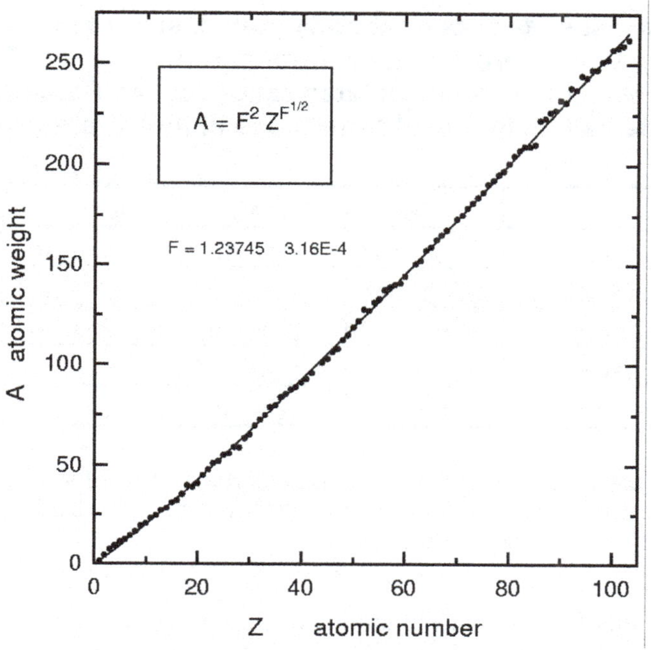


**Figure 4**. Correlation of the atomic weight *A* and atomic number *Z* for elements.

including the *Z* protons and (*A*-*Z*) neutrons dominates. Some other terms are added: correction for heat of condensation, surface tension, the number of unpaired nucleons, electrostatic repulsion, and correction due to the contribution of odd-even effect to the stability of nuclei. The relationship between *A* and *Z* for stable nuclei may be derived from the mass formula [6]. Several parameters introduced in the formula have been obtained by fitting to experimental data. *A*-*Z* correlation in the liquid drop model of nuclei is

$$Z = \frac{A}{1.98 + 0.015A^{2/3}}. \tag{20}$$

Here, the correlation of mean atomic weight *A* with atomic number *Z* is shown in Fig.4. Data are taken from the standard periodic system of elements. The longest lifetime isotopes of unstable elements with high *Z* are also taken into consideration. For our purpose, it is convenient to choose a simple function $A = aZ^b$, where *a* and *b* are parameters to be determined. The fit to experimental data is

$$A = (1.49907 \pm 0.02358)\ Z^{(1.11730 \pm 0.00357)}.$$

**Table 2. Masses $M_C$ of central body (col.1), sum of masses $m_S$ of all orbiting bodies around the central parent body (col.2), their ratios $m_S/M_C$ (col.3) and calculated ratios (col.4). Percent deviations of the values in (col.3) from those in (col.4) are given in (col.5).**

| | $M_C$/kg | $m_S$/kg | $m_S/M_C$ | $(m_S/M_C)_{calc}$ | Error % |
|---|---|---|---|---|---|
| Sun | $1.99 \cdot 10^{30}$ | $2.6756. \cdot 10^{27}$ | $1.345. \cdot 10^{-3}$ | $1.2810 \cdot 10^{-3}$ | 5 |
| Jupiter | $1.8988 \cdot 10^{27}$ | $3.9229. \cdot 10^{23}$ | $2.066 \cdot 10^{-4}$ | $2.77 \cdot 10^{-4}$ | –25 |
| Saturn | $5.685 \cdot 10^{26}$ | $1.4161 \cdot 10^{23}$ | $2.490 \cdot 10^{-4}$ | $2.13 \cdot 10^{-4}$ | 17 |
| Uranus | $8.6625 \cdot 10^{25}$ | $9.1130 \cdot 10^{21}$ | $1.052 \cdot 10^{-4}$ | $1.41 \cdot 10^{-4}$ | –25 |
| Neptune | $1.0278 \cdot 10^{26}$ | $2.1481 \cdot 10^{22}$ | $2.09 \cdot 10^{-4}$ | $1.46 \cdot 10^{-4}$ | 43 |

Note: Data are taken from: *The Astronomical Almanac for the year 1995*, Washington, U.S.Government Printing Office, p.F2,F3 and E88, and references [7] and [14] of the present paper.

Obviously, this function, at the moment, has no theoretical basis. However, the result is interesting in that parameters $a$ and $b$ can be expressed in the following way: $a = F_1^2 = (1.2244 \pm 0.0096)^2$ and $b = \sqrt{(F_2)} = (1.2484 \pm 0.0079)^{1/2}$. $F_1$ and $F_2$ are close to factor $F = 1.23$, and one may assume that the function $A(Z)$ may be written in the simplest form with only one parameter. Therefore, a new fit of the $A$-$Z$ correlation takes the form

$$A = F^2 Z^{\sqrt{F}} \qquad F = 1.2375 \pm 0.0003\,. \tag{21}$$

The plots of the functions defined by Eqs. (20) and (21) are in excellent agreement. However, the aim of the correlation presented here is not an exact determination of atomic weight, but rather an overall view on the system of the atomic nuclei.

Notes:

1. The value of $F$ in Eq.(21) is exactly to that of Eq.(5), which has resulted from the splitting of mesons and baryons (see Fig.2).
2. Eq.(21) is a power function, while in the previous equations, represented mainly by Eq.(4), functions of $F$ are exponential. The quantized values enter through the atomic numbers, Z = 1,2,3, . . . . .
3. In section 1, the factor $F$ is close to 1.23, and appears in a function with the fine-structure constant $\alpha = 0.0073$. Here, one may assume that the value 1.2375 may be presented by the sum of 1.2302 + 0.0073. The new value of $F$ = 1.2302 is close to those val-

ues in Eqs. (8), (9), (13) and (14). Therefore, one may write $A = (F + \alpha)^2 Z^{\sqrt{(F+\alpha)}}$, following the previous statement that $F$ and $\alpha$ appear together in several functions $B(F,\alpha)$.

4. The natural extension to molecular substances of the molecular weight $M$ is $M = \sum A_i = F^2 \sum (n_i Z_i^{\sqrt{F}})$, where index $i$ relates to particular atomic species.
5. Finally, one may consider Eq. (21) merely as a coincidence, due to the lack of theoretical or semi-empirical basis.

## Gravitational systems and the factor *F*

Large gravitational systems will now be considered in order to explore another aspect of the factor $F$. The model is such that there is a relationship between the sum of masses $m_S = \sum m_i$ of all orbiting bodies around a particular central body and the mass of that central body, $M_C$. For this investigation, orbital sizes, distribution of masses, interactions between bodies and many other physical processes will not be considered. It may be assumed that a mechanism which governs how much of the solar nebula would be stored in the mass of planets, or in the satellites of planets, is similar for all systems considered. This is suggested by the fact that the ratio $\sum m_i/M_C$, for all of them, is equal within an order of magnitude to $10^{-3}$ to $10^{-4}$. These data for the Sun, Jupiter, Saturn, Uranus and Neptune are given in Table 2. The mass of central bodies $M_C$ are listed in col.1, the sum $m_S$ of all orbiting planets and satellites around $M_C$ in col. 2, and their ratio $m_S/M_C$ in col.3. The correlation of $\log(m_S)$ with $\log(M_C)$ is shown in Fig.5. The best fit to observational data is given by the straight line

$$\log(m_S) = (1.220 \pm 0.040) \quad \log(M_C) + (-9.558 \pm 1.094), \tag{22}$$

or

$$m_S = 2.766 \cdot 10^{-10}\, M_C^{\,1.220}. \tag{23}$$

The calculated ratios $(m_S/M_C)_{calc}$ from the formal fit given by Eq.(23) are listed in col.4 of Table 2, and deviations (in percent) of the observational values from the calculated values are given in col.5.

By rearrangement of Eq.(23) one may obtain

$$\frac{m_S}{M_C} = \left[\frac{M_C}{2.792 \cdot 10^{43}}\right]^{0.22}. \tag{24}$$

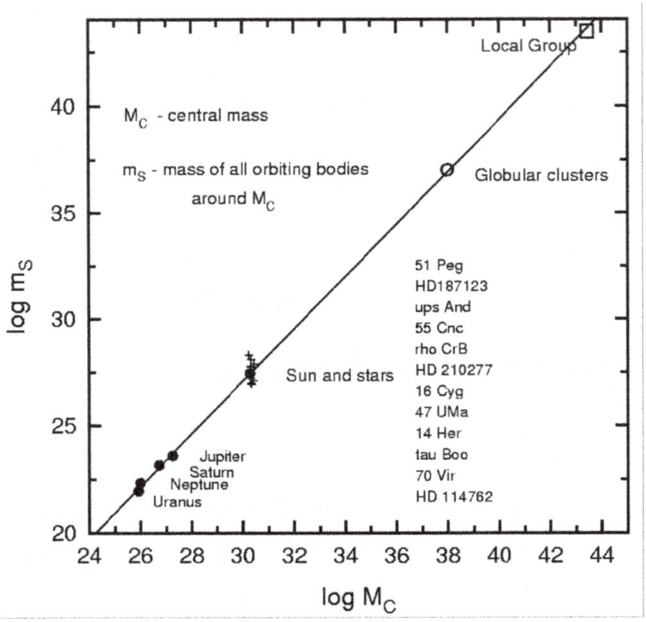


**Figure 5**. Log-log correlation of the sum $m_S$ of all orbiting bodies around the parent body of mass $M_C$, for five subsystems in the solar system (solid points). Similarly, the extra-solar planets of the stars similar to Sun are presented by crosses, globular clusters by open circle and the Local Group.

A mass of order $3 \times 10^{43}$ kg would be about 30 times the mass of M31 [7], the mass of the dominant galaxy in our Local Group. If we assume, on a trial basis, that the mass in (24) represented the mass of the Local Group (LG) then we could write

$$\frac{m_S}{M_{LG}} = \left[\frac{M_C}{M_{LG}}\right]^{1.22} \tag{25}$$

where $M_{LG} = 2.792 \cdot 10^{43}$ kg.

Of course there are many other properties of the solar system which have been recognized as quantized. Bode's law for the orbital spacing of the planets has been replaced with much more accurate fits to the data in the form of $(1.23)^n$ [2], and $r = r_1 n^2$ where $n$ is an integer, as in Bohr atomic orbits [8a,9]. The planetary masses also exhibit ratios of 1.23 [3] and even the velocities $v_n$ of the planets in orbits $n$ show a preference for values as 144 km/sec [10]. Similarly, the $nv_n$ values for planetary satellites are multiples of 24 km/sec [8b]. (The 144 km/sec and 24 km/sec are quantized values of galaxy redshifts.) There have been various suggestions in the above references as to what might be the fundamental cause for this macroscopic quantization. One suggestion might be that the matter is created in the microscopic, quantum realm and grows with time, while retaining its implanted discretization.

Owing to recently discovered extra-solar planets around stars similar to the Sun, it is possible to use available confirmed data [11] and incorporate them into Fig. 5. It is represented by crosses which are crowded around the Sun, because the masses of the stars are close to that of the Sun, and the masses of their planets are close to that of Jupiter. Thus, by including these stars, the slope of the straight line in Fig. 5 rises slightly to 1.237. If the extra-solar planets have satellites, their masses should be of the order of planetary satellites in the solar system. This would enrich the number of points in the lower part of the diagram, but only future observations will confirm or deny that expectation.

We should also note that the latest data on the extra solar planets predict accurately the observed masses and periods from the quantized formulae that solar objects obey [10,12,13].

The globular clusters are also added in Fig. 5, because they orbit the galactic centre. There are approximately $10^2$ clusters [14a], each having about $10^4$ to $10^6$ $M_o$. If an average mass for a cluster is taken to be $10^5$ $M_o$, the total mass of clusters will be roughly $10^{37}$ kg and if the mass of the centre of the galactic disk is supposed to be $10^{38}$ kg [14b], then its coordinates in Fig. 5 will be (38,37). This is represented by an open circle, but is not used in the calculations.

## Discussion and conclusion

**1**. It has been shown that the masses of elementary particles occur in discrete steps, which are related to the factor 1.23, the same factor that quantizes the values of extragalactic redshifts. Masses of elementary particles can be written in the form $m_k = m_o B^k$, $k = 0,1,2,3\ldots$ Particularly, for fundamental fermions, all masses may be expressed by the mass of the electron and some functions $B(F,\alpha)$, where $F$ is a factor close to 1.23 and $\alpha$ is the fine-structure constant. Explicit forms of $B(F,\alpha)$ are found by the trial and error method. The most accurate reproduction of current data for particle masses was that for leptons $e$, $\mu$ and $\tau$, for which the factor $F$ in two satisfactory sets of equations, Eqs.(8) and (9), is $F = 1.2292 \pm 0.0008$ and $F = 1.2311 \pm 0.0002$. For quarks, due to large uncertainties in masses, an appropriate value is $F = 1.23$. This value is also good for mesons. For baryons, however, it appears that in the simplest case $F = 1.1933$, but with an appropriate function including $\alpha$, it is again $F = 1.23$. Therefore, it seems necessary to introduce the factor $F$ if the fine-structure constant plays a major role in the scaling of particles' masses.

**2**. An increase of the atomic weight $A$ of elements with increasing atomic number $Z$ is also shown to be dependent of the factor $F$. The correlation of $A$ with $Z$ is given by $A = F^2 Z^{\sqrt{F}}$ (Eq.(21)). Clearly, there is no model supporting this formula, except that it is in excellent agreement with the $A$-$Z$ relationship derived from the liquid drop model of atomic nucleus. It is important to emphasize the difference between the rule for masses of elementary particles in the form $m_k = m_o B(F,\alpha)^k$ and the present $A$-$Z$ correlation in the form $m_k = F^2 k^{\sqrt{F}}$, where $Z$ is replaced by $k$ and $A$ by $m_k$. One can see that the first formula is an exponential function, while the second is a power function.

**3**. A power function is also found in the analysis of masses in large gravitational systems. Let $M_C$ be the mass of the central body and $m_S$ the sum of all masses orbiting the central mass. In all, there are five systems in the solar system. These are: the Sun with all planets and four planets with their satellites, *i.e.* Jupiter, Saturn, Neptune and Uranus. Correlation of $m_S$ with $M_C$ appears in the final form $m_S/M_{LG} = (M_C/M_{LG})^{1.22}$ (Eq.(25)).

**4**. Now, the great puzzle is why the factor $F$ appears in the redshift of quasars and how it is connected with elementary particles, or atomic nuclei, or with masses in gravitational systems? As mentioned in the beginning of this paper, one possible reason for the intrinsic redshift of quasars is a lower mass of electrons and other particles than their masses in our local Universe. Indeed, consider the simplest model of hydrogen atom. The light of a given wavelength $\lambda_o$ in spectrum of the hydrogen atom defined by the Bohr model (standard symbols are used) is

$$\lambda_o = \frac{2h}{(Z\alpha c)^2 m_e c\left( \frac{1}{n_1^2} - \frac{1}{n_2^2} \right)} = \frac{const.}{m_e}.$$

If it is assumed that only the mass $m_e$ of electron might be lower, due to the specific physical environment, while other quantities in the above equation are held fixed, then $\lambda_o$ becomes larger. According to the rule for the masses of elementary particles $m_k = m_o B^k$, and specifically Eqs. (7), and (17) to (19), we may suppose that the change of electron mass could be only in steps defined by the factor $F$, *i.e.*, by an equation of the type $m_{e,k} = m_e[F/\sqrt{(\alpha^0)}]^{-k} = m_e/F^k$, where fine-structure constant enters as $\alpha^0 = 1$. It follows that the wavelength $\lambda_o$ emitted by lighter hydrogen atom will be shifted to a longer value, as $\lambda_{o,k} = (const/m_e)F^k$. It comes out that the redshift $z_k$ will be equal to $(\lambda_{o,k} - \lambda_o)/\lambda_o = z_k = F^k - 1$, or $(1 + z_k) = F^k$. Obviously, $(1 + z_k) = F(1 + z_{k-1})$. So we have derived Karlsson's formula mentioned in the introduction by supposing electrons to increase their masses in steps of $F = 1.23$.

**5**. The factor $F$, besides being fundamental in fermions, also appears in composite particles such as mesons and baryons. Therefore one might suspect that in the series of atomic nuclei of elements, information about $F$ is successfully transferred. In the *A-Z* correlation, $A$ is the increasing atomic mass and $Z$ is an integer number (atomic number). Though larger than previous ones, this is still a system of the same physical character and of the same order of magnitude. What is more difficult to understand is the appearance of $F$ in large gravitational systems. This fact suggests that a similarity between microscopic and macroscopic systems might exist, but a mechanism for transferring information between these two worlds remains unknown.

## References


[1] K. G. Karlsson, *Astron.Astrophys.* **13** (1971) 333, and **58** (1997) 237.

[2] H. Arp, *Seeing Red: Redshifts, Cosmology and Academic Science*, (http://redshift.vif.com) (Montreal, Apeiron, 1998) pp.203-217.

[3] H. Arp, *Apeiron*, **2** (1995) 41.

[4] A. Czarnecki, G.P. Lepage and W.J.Marciano, *Phys. Rev*. D, 61 (2000), pp. 054313/1-6..

[5] T.F. Lee *The origin and Development of the Sun and Planets*, Melbourne 1994, ISBN 0 646 19165 9 and *Further Studies of the Solar System*, Ballarat 1997, ISBN 0 646 31840 3.

[6] E. Fermi, *Nuclear Physics* (A Course given by E.Fermi at the University of Chicago. Notes compiled by J.Orear, A.H.Rosenfeld and R.A.Schluter), The University of Chicago Press, 1951, p.7.

[7] M. Zeilik and J. Gaustad, *Astronomy-The Cosmic Perspective*, J.Wiley Sons, Inc., New York, 1990, p.665.

[8] A. Rubčić and J. Rubčić, a) *Fizika B* **4** (1995) 11, b) *Fizika B* **7** (1998) 1.

[9] A. G. Agnese and R. Festa, *Phys.Lett. A* **227** (1997) 165.

[10] L. Nottale, *Astron. Astrophys*, **315** (1996) L9.

[11] J. Schneider, *Extrasolar Planets Encyclopaedia*, http//www.obspm.fr/planets.

[12] A. Rubčić and J. Rubčić, *Fizika B* **5** (1996) 85 and *Fizika A* **8** (1999) 45.

[13] A. G. Agnese and R. Festa, astro-ph/9910534 (1999).

[14] H. Karttunen, P. Kröger, H. Oja, M. Poutanen, and K.J. Donner (Editors), *Fundamental Astronomy*, New York, Springer, 1996, a) p.384, b) p.408.

# Testing the Hypothesis of Redshift Quantization in Iwanowska Galaxy Lines Connected with our Galaxy and M31

Konrad Rudnicki[13]
Włodzimierz Godłowski[14]
Anna Magdziarz[15]

A search for quantization of redshifts was performed for objects belonging to the Iwanowska lines connected with our Galaxy and M31. A distinct periodisation effect was found. No exact quantization effect was detected. The accuracy of observational data was not sufficient for more precise analysis. No correlation between values of redshift or morphological types of objects and locations on lines was visible.

*Keywords*: redshift quantization, globular cluster

## 1. Introduction

The existence of straight bipolar lines of extragalactic objects connected with our Galaxy and with M31 was postulated by Wihelmina Iwanowska (1989). She considered not only galaxies but also some globular clusters as belonging to those lines. The aim of the present paper is to check whether the objects in Iwanowska Lines reveal the redshift periodization or quantization, and whether globular clusters reveal similar regularities of spectral shifts as galaxies.

## 2. Data

A total of 40 galaxies were considered: all galaxies from the original paper of Iwanowska having the necessary data, and also all globular clusters from this paper. Iwanowska made this selection according to the positions of objects on the celestial sphere only. We were unable to check their spatial alignment due to the absence of photometric distances to many of them. The redshifts

[13] Astronomical Observatory of the Jagiellonian University, Cracow
[14] Astronomical Observatory of the Jagiellonian University, Cracow
[15] Institute of Mathematics and Physics, Warmian and Mazurian University, Olsztyn

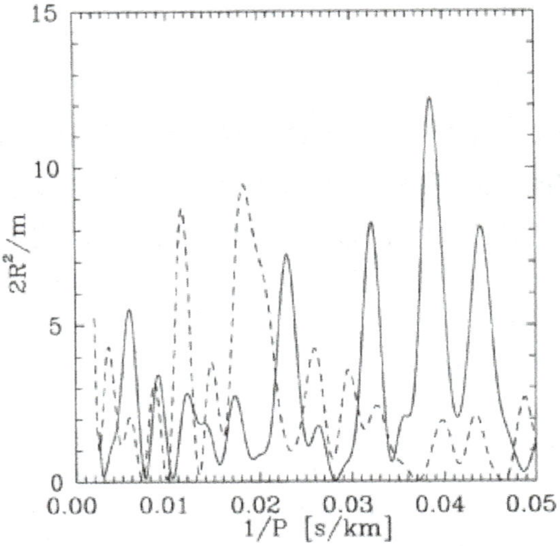


**Fig.1**. Power spectrum analysis for all galaxies, with known spectral shifts, belonging to Iwanowska lines—solid line. The same for a random uniform distribution—dashed line.

of the selected objects were taken from the most reliable existing sources, different for different objects. 6 additional galaxies, members of the local group, possessing known redshifts but not mentioned by Iwanowska, were added: see below (Tables 1-3).

The redshifts were corrected for the galactic orbital motion of the Sun (reduced to the reference system of the galactic centre) according to the two assumptions of Guthrie and Napier (1991) X = 0, Y = 220, $Z = 0$ km/sec ($\alpha$) and $X = -12$, $Y = 233$, $Z = 8$ km/sec ($\beta$) and Guthrie and Napier (1996) $X = -11$, $Y = 213$, $Z = 7$ km/sec ($\gamma$). Calculations with the standard solar vector $X = -33$, $Y = 232$, $Z = 8$ km/sec, and with pure suncentric redshifts $X = 0$, $Y = 0$, $Z = 0$ were also performed. In the following, only results obtained with Guthrie and Napier's 1996 corrected redshifts are presented. The effects with other solar vector data give ambiguous results and were used here for comparison only.

## 3. Methods of analysis

All calculations were performed for all the objects together, for all galaxies and all globular clusters separately, as well as galaxies together with globular clusters for individual Iwanowska lines.

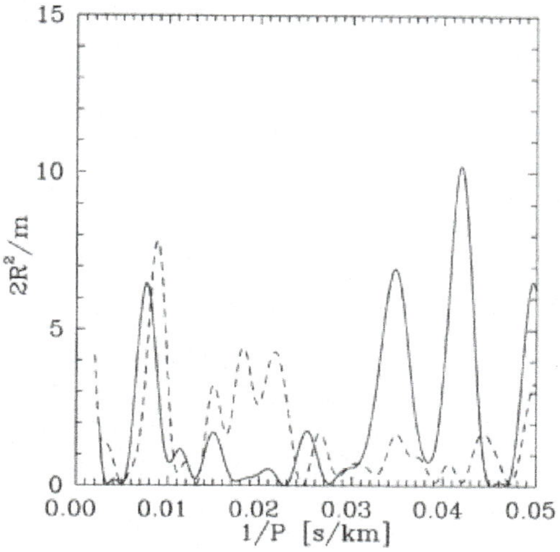


**Fig.2**. Power spectrum analysis for the line Galaxy A. Solid and dashed lines as on Fig.1.

### 3.1 The analysis of mean errors

We tried to obtain and to visualise the results in the simplest way, as follows. All the spectral shifts were divided into bins of 36 km/sec with the initial bin around $cz = 0$ according to Tifft and Cocke (1984). In every bin the actual mean value of redshifts was calculated without weights (weights introduce bias when various objects are measured, not the same one) and the actual dispersion of redshifts $S^*$ was calculated. The theoretical mean error $m$ of mean redshift according to the definition $m = [\sum m_i/n]^{-2}$ (Zonn 1952, compare also Bronsteyn and Semendyaev 1957) was calculated from mean errors $m_i$ given by the authors of the measurements. The value $cz = 0$ and the mean error of the established solar orbital velocity as $m_i$ were accepted for our Galaxy. The values $S^*$ were compared with the values of $m$ (Table 4). The values of $S^*$ are of the same order of value as $m$.

The next step was to test the hypothesis that the redshifts of our sample of galaxies are strictly quantized, and any deviation from a discrete distribution of redshifts is only a result of measurement errors. If that hypothesis is true then the value

$$M = \Sigma \frac{\left(V_{obs,i} - V_{q,i}\right)^2}{m_i^2}$$

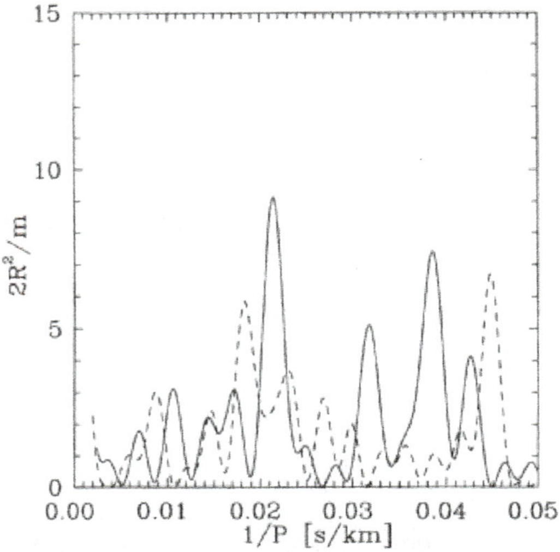


**Fig. 3**. Power spectrum analysis for the line M31C. Solid and dashed lines as on Fig. 1.

is distributed as $\chi^2$ with $n$ degrees of freedom (compare Brandt 1970) where $V_q$ is redshift according to the quantization hypothesis, $V_{obs}$—the observed redshift, and $m_i$ is the mean error of redshift measurements. The sum is extended over all objects involved in a given calculation. This test excludes the pure quantization hypothesis on the 0.99 significance level for 1 km/sec accuracy of measurements. This result is the same when the values $k{\cdot}36$ km/sec as $V_q$ were accepted ($k = ...-2,-1,0,1,2,...$) or actual mean values were calculated.

## 3.2 The power spectrum analysis

The power spectrum analysis in the form of Webster (1976) using the Rayley test was performed. This method is very adequate for searching for any periodicity in the class of irregular distributed points. We used it to test the hypothesis that the quantization has a period of 36 km/sec, as well as to look for the best fitting period. The results obtained are given in Table 5 and on Fig. 1-3.

We found, using this method, that all the galaxies show a periodization effect when taken together (significance level 0.96), as well as the *Galaxy A* and *M31 C* lines (significance level 0.94). Adding the galaxies belonging to the Local Group but not connected with Iwanowska lines (see Table 1) to the

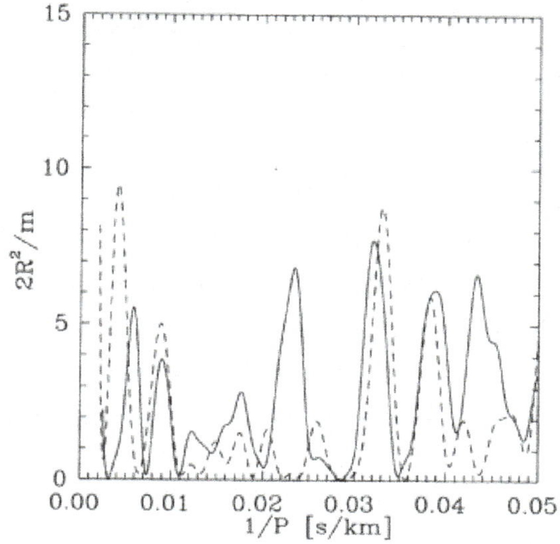


**Fig.4**. Power spectrum analysis for all galaxies with known spectral shifts from Table 1. Solid and dashed lines as on Fig.1.

sample destroys the periodization effect. No effect is visible for the pure sample of globular clusters either.

## 4. Results

Periodization of redshifts for all galaxies belonging to the Iwanowska lines has been confirmed at a 0.95 significance level. The same was found for *Galaxy A* and *M31 C* lines (galaxies and globular star clusters together). No effect was found for other lines or for the sample of all the globular clusters taken separately. No strict quantization effect was found. Allowance must, however, be made for the fact that many theoretical values of $m$ calculated from the actual observational errors are exceedingly high, sometimes exceeding even the value $\delta = 10$ km/sec characteristic for a random uniform distribution of redshifts (see Table 4). This shows that the existing observational data are in principle not accurate enough to study the subtle quantization structure in detail. The same follows from the fact that some values $m$ are higher than $S^*$.

During this work, we also looked for any possible correlation between the localization of objects on a given line and value of redshifts, as well as for a correlation between morphological types of objects and redshifts. None were found.

We wish to express our gratitude to the late Paweł Magdziarz for important discussions and help in some theoretical and numerical problems.

## References

Brandt, S. 1970 *Statistical and Computational Methods in Data Analysis* New York.

Bronsteyn, I.N. & Semendyaev, K.A. 1957 *Spravocnik po Matematike* Moskva.

Guthrie, B.N.G. & Napier, W.M. 1991 *Month.Notic.Roy.Astr.Soc* **253**, 533.

Guthrie, B.N.G. & Napier, W.M. 1996 *Astr.Astroph* **310**, 353.

Irwin, M. 2000 http://www.ast.cam.ac.uk/~mike/local_members.html.

Iwanowska, W. 1989 in: *From Stars to Quasars* Ed. Grudzinska and Krygier, Torun p.55.

Tifft W.G. & Cocke, W.J. 1984 *Astroph.Journ.* **287**, 492.

Webster, A. 1976 *Month.Notic.Roy.Astr.Soc* **175**, 61.

Zonn, W. 1952 *O Błędach Obserwacyjnych i Ich Wyrównywaniu* Warszawa.

# Table 1

## Galaxies belonging to Iwanowska lines; galaxies belonging to the Local Group according to Irvin are added

| Nr | PCG | NGC | name | morph. | $\alpha_{1950.0}$ | $\delta_{1950.0}$ | l | b | $v_{21}$ | $v_{opt}$ | $v_{Irwin}$ |
|---|---|---|---|---|---|---|---|---|---|---|---|
| 1 | 1305 | | IC 10 | Ir IV | 00 17 42 | + 59 00 52 | 118.97 | −03.34 | −344 ± 3 | −342 ± 22 | −344 ± 4 |
| 2 | | | Cetus | dE4 | 00 23 36 | −11 19 | 101.40 | −72.80 | | | |
| 3 | 2004 | 147 | DDO 3 | dE5 | 00 30 27 | + 48 13 56 | 119.82 | −14.25 | | −160 ± 18 | −157± 18 |
| 4 | 2121 | | And III | Dsph | 00 32 42 | + 36 14 | 119.34 | −26.25 | | | |
| 5 | 2329 | 185 | | dE3p | 00 36 12 | + 48 03 50 | 120.79 | −14.48 | | −251 ± 13 | −208± 13 |
| 6 | 2555 | 221 | M32 | E2 | 00 36 58 | + 40 35 29 | 121.15 | −21.98 | | −205 ± 8 | −190± 8 |
| 7 | 2429 | 205 | | E6p | 00 37 39 | + 41 24 44 | 120.72 | −21.14 | −232±10 | −254 ± 14 | −239± 12 |
| 8 | 2557 | 224 | M31 | Sb I-II | 00 40 00 | + 40 59 43 | 121.17 | −21.57 | −300 ± 4 | −295 ± 7 | −299± 5 |
| 9 | 2666 | | And I | Dsph | 00 42 48 | + 37 46 | 121.65 | −24.82 | | | |
| 10 | 3085 | | SMC | Ir IV-V | 00 50 53 | −73 04 18 | 302.81 | −44.33 | + 175 ± 7 | + 190 ± 49 | + 163±10 |
| 11 | 3589 | | Scl | Dsph | 00 57 47 | −33 58 42 | 287.53 | −83.16 | | + 148 ± 22 | + 107±22 |
| 12 | 3792 | | LGS 3 | DIr | 01 01 12 | + 21 37 | 126.75 | −40.89 | −279 ± 5[1] | | −277 ± 5 |
| 13 | 3844 | | IC 1613 | Ir V | 01 02 20 | + 01 51 56 | 129.79 | −60.56 | −234 ± 1[2] | −237 ± 22 | −236 ± 1 |
| 14 | 4126 | 404 | | S0 | 01 06 39 | + 35 27 06 | 127.03 | −27.01 | −48 ± 9 | −19 ± 29 | −45 ± 12 |
| 15 | | | AND V | dE | 01 07 18 | + 47 22 | 126.20 | −15.10 | | | |
| 16 | 4601 | | And II | Dsph | 01 13 36 | + 33 11 | 128.89 | −29.14 | | | |
| 17 | 5818 | 598 | M33 | Sc II-III | 01 31 02 | + 30 24 15 | 133.61 | −31.33 | −179 ± 3 | −204 ± 17 | −180 ± 4 |
| 18 | | | Phoenix | Ir | 01 49 00 | −44 42 00 | 272.19 | −68.95 | | | + 56± 10 |
| 19 | 9892 | | Maffei 1 | E | 02 32 36 | + 59 26 | 135.83 | −00.57 | | + 2 ± 72 | + 2 ± 72 |
| 20 | 10093 | | Fornax | Dsph | 02 37 55 | −34 39 48 | 237.10 | −65.65 | + 53 ± 9[3] | + 47 ± 34 | + 53± 12 |
| 21 | 10217 | | Maffei 2 | Sbc | 02 38 08 | + 59 23 24 | 136.50 | −00.33 | −1 ± 6 | | −1 ± 6 |
| 22 | 15345 | 1569 | | Ir | 04 26 06 | + 64 44 18 | 143.68 | + 11.24 | −98 ± 4 | −74 ± 17 | −77 ± 6 |
| 23 | 15488 | 1560 | | Sd | 04 27 08 | + 71 46 29 | 138.37 | + 16.02 | −36 ± 5 | −194 ± 73 | −40 ± 7 |
| 24 | 15439 | | UGCA 92 | Ir | 04 27 24 | + 63 30 | 144.71 | + 10.51 | −99 ± 5 | | −103 ± 5 |
| 25 | | | Cam A | Ir | 04 31 30 | + 71 25 00 | 138.88 | + 16.04 | | | |
| 26 | 17223 | | LMC | Ir III-IV | 05 24 00 | −69 48 00 | 280.47 | −32.89 | + 324±10 | + 227 ± 19 | + 270±13 |
| 27 | 19441 | | Carina | Dsph | 06 40 24 | −50 55 00 | 260.11 | −22.22 | | + 229 ± 60 | + 223±60 |
| 28 | | | ARGO | Ir | 07 04 30 | −58 27 00 | 268.96 | −21.15 | | | + 554±10 |
| 29 | 21600 | | DDO 47 | Ir | 07 39 00 | + 16 55 | 203.10 | + 18.54 | + 270 ± 4 | | + 270 ± 4 |
| 30 | 28868 | | Leo A | Ir V | 09 56 24 | + 30 59 | 196.90 | + 52.41 | + 20 ± 4 | | + 26 ± 4 |
| 31 | 28913 | | Sextans B | Ir | 09 57 23 | + 05 34 22 | 233.20 | + 43.78 | + 301 ± 4 | | + 301 ± 4 |
| 32 | 29128 | 3109 | DDO 236 | Ir | 10 00 49 | −25 55 00 | 262.10 | + 23.07 | + 403 ± 1[1] | + 408 ± 58 | + 403 ±1 |
| 33 | | | Antlia | dE3 | 10 01 48 | −27 05 00 | 263.10 | + 22.32 | | | 361± 10 |
| 34 | 29488 | | Leo I | Dsph | 10 05 47 | + 12 33 10 | 225.98 | + 49.11 | | + 168 ± 60 | + 285±60 |
| 35 | 29653 | | Sextans A | Ir | 10 08 30 | −04 28 00 | 246.17 | + 39.86 | + 324 ± 4 | + 374 ± 19 | + 325 ± 6 |
| 36 | | | Sextans | dE4 | 10 10 42 | −01 24 00 | 243.55 | + 42.27 | | | + 224± 0 |
| 37 | 34176 | | Leo II | Dsph | 11 10 50 | + 22 25 32 | 220.17 | + 67.23 | | + 90 ± 60 | + 76± 60 |
| 38 | 35286 | | UGC 6456 | P | 11 24 36 | + 79 16 | 127.84 | + 37.33 | −93 ± 5 | −49 ± 32 | −92 ± 7 |
| 39 | 39346 | 4236 | | Sdm | 12 14 22 | + 69 44 36 | 127.43 | + 47.36 | 0 ± 4 | −5 ± 58 | 0 ± 6 |
| 40 | 44491 | | Gr 8 | Ir | 12 56 06 | + 14 29 | 310.72 | + 76.98 | + 214 ± 3 | + 242 ± 35 | + 216 ± 4 |
| 41 | 50961 | | DDO 187 | Ir | 14 13 36 | + 23 17 | 25.57 | + 70.47 | + 154 ± 4 | | + 154 ± 4 |
| 42 | 54074 | | UMi | Dsph | 15 08 12 | + 67 23 | 104.97 | + 44.84 | | −209 ± 24 | −250 ±24 |
| 43 | 60095 | | Draco | Dsph | 17 19 24 | + 57 57 50 | 86.37 | + 34.72 | | −281 ± 24 | −289 ±24 |
| 44 | | | Galaxy | Sbc | 17 42 24 | −28 55 50 | 0.00 | + 00.00 | | | 0 ± 0 |
| 45 | | | Sagittar | dE7 | 18 51 54 | −30 30 00 | 5.65 | −14.08 | | | + 140±10 |
| 46 | 63287 | | Sgr | Ir | 19 27 06 | −17 47 00 | 21.06 | −16.29 | −77 ± 4 | | −79 ± 4 |
| 47 | 63613 | 6822 | DDO 209 | Ir IV–V | 19 42 08 | −14 55 29 | 25.34 | −18.40 | −56 ± 4 | −26 ± 20 | −49 ± 6 |
| 48 | 65367 | | Aqr | Ir | 20 44 06 | −13 02 | 34.05 | −31.35 | −137 ± 5 | | −131± 5 |
| 49 | 67908 | | IC 5152 | Ir IV | 21 59 26 | −51 32 18 | 343.92 | −50.19 | + 123 ± 4[3] | + 92 ± 41 | + 121± 6 |
| 50 | | | Tucana | dE5 | 22 38 30 | −64 41 00 | 322.90 | −47.37 | | | |
| 51 | | | UKS2323- | Ir | 23 23 48 | −32 40 00 | 11.86 | −70.86 | | | + 62 ±10 |
| 52 | | | AND VII | dE3 | 23 24 12 | + 30 25 | 109.50 | −09.90 | | | |
| 53 | 71538 | | Peg Ir | Ir V | 23 26 03 | + 14 28 16 | 94.77 | −43.55 | −183 ± 2[2] | | −181 ± 2 |
| 54 | | | AND VI | dE3 | 23 49 12 | + 24 18 | 106.00 | −36.30 | | | |
| 55 | 143 | | WLM | Ir IV-V | 23 59 23 | −15 43 43 | 75.87 | −73.61 | −117 ± 3[5] | −75 ± 21 | −116 ± 4 |

[1]—Lo K.Y., *et al.*, 1993, *AJ*, 106, 507
[2]—Lu N.Y., *et al.*, 1993, *ApJS*, 88, 383
[3]—1989, *ESOLV.C*
[4]—Schneider S.E., *et al.*, 1992, *ApJS*, 81, 5
[5]—Huchra J., *et al.*, 1993, *AJ*, 105, 1637

## Table 2

### Globular star clusters belonging to Iwanowska lines

| Nr | Nazwa | $\alpha_{1950.0}$ | $\delta_{1950.0}$ | l | b | $v_{opt}$ | $v_\alpha$ | $v_\beta$ | $v_\gamma$ |
|---|---|---|---|---|---|---|---|---|---|
| 1 | Pal 1 | 03 26 04 | + 79 24 39 | 130.07 | + 19.03 | + 3 ±32[3] | + 168±32 | + 181±32 | + 166±32 |
| 2 | AM 1 | 03 53 35 | –49 45 36 | 258.36 | –48.47 | + 116 ± 15[2] | –41±15 | –40±15 | –26±15 |
| 3 | Eri | 04 22 35 | –21 18 06 | 218.11 | –41.33 | –21 ± 4[4] | –138± 4 | –127±4 | –118±4 |
| 4 | NGC 2419 | 07 34 46 | + 38 59 44 | 180.37 | + 25.24 | –20 ± 5[1] | –27± 5 | –7±5 | –8±5 |
| 5 | Pal 3 | 10 02 57 | + 00 18 54 | 240.14 | + 41.86 | + 89 ± 9[1] | –59 ± 9 | –52±9 | –40±9 |
| 6 | Pal 4 | 11 26 38 | + 29 14 57 | 202.31 | + 71.80 | + 75 ± 5[1] | + 52 ± 5 | + 58±5 | + 60±5 |
| 7 | Pal 14 | 16 08 47 | + 15 05 12 | 28.76 | + 42.18 | + 72 ± 4[4] | + 165 ± 4 | + 153±4 | + 146± 4 |
| 8 | NGC 6229 | 16 45 34 | + 47 36 57 | 73.64 | + 40.31 | –154 ± 8[3] | + 22 ± 8 | + 19±8 | + 4±8 |
| 9 | Pal 8 | 18 38 32 | –19 52 30 | 14.10 | –06.80 | –38 ± 20[5] | + 26 ± 20 | + 6±20 | + 2±20 |
| 10 | NGC 7006 | 20 59 09 | + 15 59 25 | 63.77 | –19.41 | –385 ± 7[3] | –187 ± 7 | –195±7 | –212±7 |
| 11 | Pal 13 | 23 04 14 | + 12 30 05 | 87.10 | –42.70 | –28 ± 37[3] | + 138± 37 | + 137±37 | + 123±37 |

[1]—Olszewski E.W., *et al.*, 1986, *ApJ*, 302, L45
[2]—Suntzeff N., *et al.*, 1985, *AJ*, 90, 1481
[3]—Webbink R.F., 1981, *ApJS*, 45, 259
[4]—Zaritsky D., *et al.*, 1989, *ApJ*, 345, 759
[5]—Zinn R., West M.J., 1984, *ApJS*, 55, 45

## Table. 3

### Membership of Iwanowska's lines

| | |
|---|---|
| Galaxy A | N Umi, Dra, NGC6229, Pal1, Pal14, NGC4236, NGC6456, S LMC, SMC, Car, Eri, AM1 |
| Galaxy B | N Pal3, Pal4, LeoI, Leo II, LeoA, SexA, SexB, NGC 2419, NGC 3109, GR8, DDO187, DDO47 S Pal 8, Pal 13, NGC 7006, 6822, Sc1, For, Aqr, Sgr, WLM, IC5152 |
| M31 A | N NGC185, NGC 147, IC10 S M32, AndI, AndIII, LGS 3, I1613 |
| M31 B | N NGC 205 S And II, M33, NGC 404 |
| M31 C | N Maffei 1, Maffei 2, NGC1560, NGC 1569, A92 S Peg |

## Table 4

### The analysis of mean observational errors, case $\gamma$. Value of $k$ (column 1), number of object (column 2), theoretical Tifft values (column 3), calculated mean spectral shifts in bins (column 4), dispersion $S^*$ (column 5) and mean error of observations (column 6).

#### Galaxies + Globular Star Clusters

| | | | | | |
|---|---|---|---|---|---|
| –6 | 1 | $v_{tif}$ = –216. | $v_s$ = –212. | $s^*$ = ? | $m$ = 7.0 |
| –5 | 0 | $v_{tif}$ = –180. | $v_s$ = 0. | $s^*$ = 0.0 | $m$ = 0.0 |
| –4 | 4 | $v_{tif}$ = –144. | $v_s$ = –146. | $s^*$ = 13.8 | $m$ = 4.1 |
| –3 | 3 | $v_{tif}$ = –108. | $v_s$ = –109. | $s^*$ = 10.6 | $m$ = 19.7 |
| –2 | 2 | $v_{tif}$ = –72. | $v_s$ = –65. | $s^*$ = 0.2 | $m$ = 8.9 |
| –1 | 8 | $v_{tif}$ = 36. | $v_s$ = –32. | $s^*$ = 9.3 | $m$ = 9.5 |
| 0 | 6 | $v_{tif}$ = 0. | $v_s$ = –3. | $s^*$ = 8.1 | $m$ = 10.2 |
| 1 | 7 | $v_{tif}$ = 36. | $v_s$ = 31. | $s^*$ = 9.5 | $m$ = 33.2 |
| 2 | 5 | $v_{tif}$ = 72. | $v_s$ = 70. | $s^*$ = 12.9 | $m$ = 12.2 |
| 3 | 4 | $v_{tif}$ = 108. | $v_s$ = 116. | $s^*$ = 9.5 | $m$ = 20.0 |
| 4 | 3 | $v_{tif}$ = 144. | $v_s$ = 152. | $s^*$ = 6.4 | $m$ = 41.8 |
| 5 | 6 | $v_{tif}$ = 180. | $v_s$ = 184. | $s^*$ = 9.7 | $m$ = 28.0 |
| 6 | 2 | $v_{tif}$ = 216. | $v_s$ = 208. | $s^*$ = 7.4 | $m$ = 2.9 |

**Galaxies**

| | | | | | |
|---|---|---|---|---|---|
| –4 | 4 | $v_{tif}$ = –144. | $v_s$ = –146. | $s^*$ = 13.8 | $m$ = 4.1 |
| –3 | 2 | $v_{tif}$ = –108. | $v_s$ = –104. | $s^*$ = 9.7 | $m$ = 24.0 |
| –2 | 2 | $v_{tif}$ = –72. | $v_s$ = 65. | $s^*$ = 0.2 | $m$ = 8.9 |
| –1 | 6 | $v_{tif}$ = –36. | $v_s$ = –31. | $s^*$ = 10.1 | $m$ = 8.4 |
| 0 | 3 | $v_{tif}$ = 0. | $v_s$ = –6. | $s^*$ = 10.0 | $m$ = 6.6 |
| 1 | 7 | $v_{tif}$ = 36. | $v_s$ = 31. | $s^*$ = 9.5 | $m$ = 33.2 |
| 2 | 4 | $v_{tif}$ = 72. | $v_s$ = 73. | $s^*$ = 13.2 | $m$ = 13.5 |
| 3 | 3 | $v_{tif}$ = 108. | $v_s$ = 113. | $s^*$ = 9.8 | $m$ = 8.7 |
| 4 | 2 | $v_{tif}$ = 144. | $v_s$ = 156. | $s^*$ = 3.4 | $m$ = 51.1 |
| 5 | 5 | $v_{tif}$ = 180. | $v_s$ = 188. | $s^*$ = 4.5 | $m$ = 27.1 |
| 6 | 2 | $v_{tif}$ = 216. | $v_s$ = 208. | $s^*$ = 7.4 | $m$ = 2.9 |

**Galaxies + Globular Star Clusters + missing LG Galaxies**

| | | | | | |
|---|---|---|---|---|---|
| –6 | 1 | $v_{tif}$ = –216. | $v_s$ = –212. | $s^*$ = ? | $m$ = 7.0 |
| –5 | 0 | $v_{tif}$ = –180. | $v_s$ = 0. | $s^*$ = 0.0 | $m$ = 0.0 |
| –4 | 4 | $v_{tif}$ = –144. | $v_s$ = –146. | $s^*$ = 13.8 | $m$ = 4.1 |
| –3 | 3 | $v_{tif}$ = –108. | $v_s$ = –109. | $s^*$ = 10.6 | $m$ = 19.7 |
| –2 | 2 | $v_{tif}$ = –72. | $v_s$ = –65. | $s^*$ = 0.2 | $m$ = 8.9 |
| –1 | 9 | $v_{tif}$ = –36. | $v_s$ = –31. | $s^*$ = 8.9 | $m$ = 9.6 |
| 0 | 6 | $v_{tif}$ = 0. | $v_s$ = –3. | $s^*$ = 8.1 | $m$ = 10.2 |
| 1 | 7 | $v_{tif}$ = 36. | $v_s$ = 31. | $s^*$ = 9.5 | $m$ = 33.2 |
| 2 | 6 | $v_{tif}$ = 72. | $v_s$ = 70. | $s^*$ = 11.7 | $m$ = 11.9 |
| 3 | 5 | $v_{tif}$ = 108. | $v_s$ = 111. | $s^*$ = 13.6 | $m$ = 18.4 |
| 4 | 4 | $v_{tif}$ = 144. | $v_s$ = 151. | $s^*$ = 5.7 | $m$ = 36.5 |
| 5 | 7 | $v_{tif}$ = 180. | $v_s$ = 182. | $s^*$ = 10.5 | $m$ = 26.2 |
| 6 | 2 | $v_{tif}$ = 216. | $v_s$ = 208. | $s^*$ = 7.4 | $m$ = 2.9 |
| 7 | 0 | $v_{tif}$ = 252. | $v_s$ = 0. | $s^*$ = 0.0 | $m$ = 0.0 |
| 8 | 0 | $v_{tif}$ = 288. | $v_s$ = 0. | $s^*$ = 0.0 | $m$ = 0.0 |
| 9 | 0 | $v_{tif}$ = 324. | $v_s$ = 0. | $s^*$ = 0.0 | $m$ = 0.0 |
| 10 | 1 | $v_{tif}$ = 360. | $v_s$ = 353. | $s^*$ = ? | $m$ = 10.0 |

**Galaxies + missing LG Galaxies**

| | | | | | |
|---|---|---|---|---|---|
| –4 | 4 | $v_{tif}$ = –144. | $v_s$ = –146. | $s^*$ = 13.8 | $m$ = 4.1 |
| –3 | 2 | $v_{tif}$ = –108. | $v_s$ = –104. | $s^*$ = 9.7 | $m$ = 24.0 |
| –2 | 2 | $v_{tif}$ = –72. | $v_s$ = –65. | $s^*$ = 0.2 | $m$ = 8.9 |
| –1 | 7 | $v_{tif}$ = –36. | $v_s$ = –31. | $s^*$ = 9.4 | $m$ = 8.6 |
| 0 | 3 | $v_{tif}$ = 0. | $v_s$ = –6. | $s^*$ = 10.0 | $m$ = 6.6 |
| 1 | 7 | $v_{tif}$ = 36. | $v_s$ = 31. | $s^*$ = 9.5 | $m$ = 33.2 |
| 2 | 5 | $v_{tif}$ = 72. | $v_s$ = 72. | $s^*$ = 11.8 | $m$ = 12.8 |
| 3 | 4 | $v_{tif}$ = 108. | $v_s$ = 108. | $s^*$ = 13.5 | $m$ = 9.1 |
| 4 | 3 | $v_{tif}$ = 144. | $v_s$ = 153. | $s^*$ = 5.1 | $m$ = 42.1 |
| 5 | 6 | $v_{tif}$ = 180. | $v_s$ = 185. | $s^*$ = 8.5 | $m$ = 25.1 |
| 6 | 2 | $v_{tif}$ = 216. | $v_s$ = 208. | $s^*$ = 7.4 | $m$ = 2.9 |
| 7 | 0 | $v_{tif}$ = 252. | $v_s$ = 0. | $s^*$ = 0.0 | $m$ = 0.0 |
| 8 | 0 | $v_{tif}$ = 288. | $v_s$ = 0. | $s^*$ = 0.0 | $m$ = 0.0 |
| 9 | 0 | $v_{tif}$ = 324. | $v_s$ = 0. | $s^*$ = 0.0 | $m$ = 0.0 |
| 10 | 1 | $v_{tif}$ = 360. | $v_s$ = 353. | $s^*$ = ? | $m$ = 10.0 |

## Table 5

**Period of the highest peak from the Rayley test (column 2) and probability that this peak is produced from random uniform distribution (column 3)**

| Sample | *V* | *P(I)* |
|---|---|---|
| Galaxies + globular clusters | 30.8 | 0.105 |
| Galaxies | 25.9 | 0.040 |
| Globular Clusters | 3.8 | 0.774 |
| Galaxy A | 23.8 | 0.054 |
| Galaxy B | 63.5 | 0.267 |
| M31 A | 25.7 | 0.363 |
| M31 B | 21.6 | 0.349 |
| M31 C | 46.8 | 0.065 |
| Galaxies A (only galaxies) | 32.1 | 0.360 |
| Galaxies B (only galaxies) | 32.3 | 0.209 |
| | | |
| Galaxies + globular clusters + missing LG Galaxies | 30.6 | 0.162 |
| Galaxies + missing LG Galaxies | 31.2 | 0.351 |

# Application of Thermodynamics to Cosmology

Bernard R. Bligh

When cosmologists say that the Universe was at a temperature of $10^{10}$ K when it was one second young, they are making a *thermodynamic* statement. Temperature is a thermodynamic function. Cosmologists frequently apply thermodynamic equations wrongly and they have the erroneous belief that an expanding gas automatically cools down. The Hot Big Bang Theory stands or falls on its thermodynamic credentials. Cosmologists have entered the realm of Thermodynamics.

This paper analyses some of the thermodynamic arguments leading to the Hot Big Bang Theory and shows that the theory does not exhibit a valid energy balance.

Thermodynamic calculations are presented with the aid of a Temperature-Entropy Diagram for hydrogen and these calculations together with some graphs show conclusively that the Hot Big Bang Theory violates both the First and Second Laws of Thermodynamics.

An alternative version of the Hot Big Bang Theory—that the cooling of the Universe comes about by expanding against the force of gravity—is disproved by a theorem using the Thermodynamics of an isentropic expansion.

Proponents of the Hot Big Bang Theory are challenged to answer five questions.

## Nomenclature

$Cp$ = Specific heat capacity at constant pressure, J mole-1 $K^{-1}$,

$Cv$ = Specific heat capacity at constant volume, J $mole^{-1}$ $K^{-1}$,

$H$ = Enthalpy, J $mole^{-1}$,

$h$ = Planck's constant, = $6.6256 \times 10^{-34}$ J s,

$k$ = Boltzmann's constant, = $1.3805 \times 10^{-23}$ J $K^{-1}$,

$M$ = Relative Molecular Mass (RMM) molecular weight, dimensionless,

$m$ = Mass of a molecule, kg,

$n$ = Number of moles in a system,

$NA$ = Avogadro number, *i.e.* the number of molecules in a mole,
$= 6.023 \times 10^{26}$ number kg –mole$^{-1}$,
$N$ = Number of molecules in a system, $= n \times NA$,
$P$ = Pressure, $Pa = N\, m^{-2}$,
$Q$ = A Quantity of heat, J,
$R$ = Gas constant, = 8314 J mole$^{-1}$ K$^{-1}$,
$R$ = Scaling factor in Cosmology,
$S$ = Entropy, J mole$^{-1}$ K$^{-1}$,
$T$ = Absolute temperature, K,
$U$ = Internal energy, J mole$^{-1}$,
$V$ = Volume, m$^3$,
$W$ = Work, J,
$\gamma$ = Ratio of specific heats, $= Cp/Cv$, dimensionless.

## Introduction

Let us start with the First Law of Thermodynamics which says that if a quantity of heat, $Q$, is put into a system, either the internal energy is increased by $\Delta U$ or some work, $W$, is obtained or there is a combination of both, *i.e.*,

$$Q = \Delta U + W. \tag{1}$$

Note that $W$ does not need to be mechanical work, it could, for example, be electrical work. If W does represent a machine such as a turbine, the First Law of Thermodynamics makes no stipulation as to how efficient that machine must be.

Now let us look at another equation,

$$Q = \Delta U + P\Delta V. \tag{2}$$

Contrary to what appears in books on Cosmology, this equation is not a general statement of the First Law of Thermodynamics. The reader can test this for himself/herself.

Consider the action of letting air out of a car tyre. If the tyre is at atmospheric temperature, no heat goes in or out of the system, so the action is adiabatic, *i.e.*,

$$Q = 0.$$

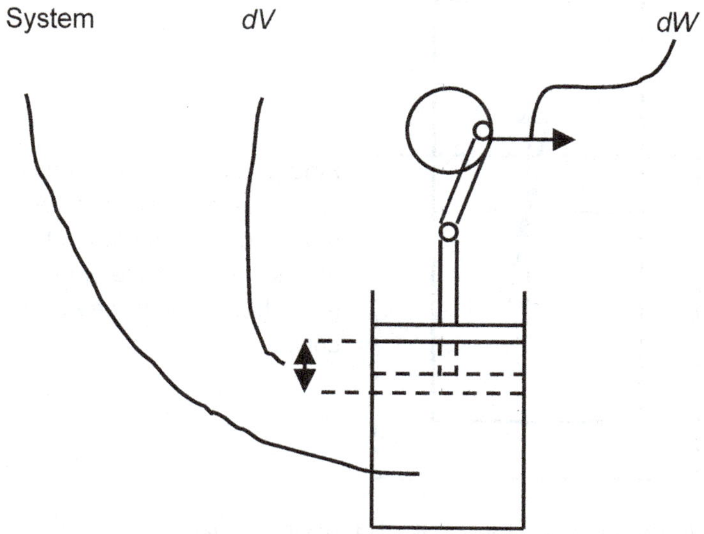


**Figure 1**. Removal of work *dW* is external to the system. If the cylinder is adiabatic, dQ = 0 and $dU + PdV = 0$

Now air is virtually an ideal gas and it is a property of an ideal gas that the internal energy, $U$, is a constant at constant temperature and is independent of the volume. Therefore $\Delta U = 0$.

Even if the experimenter did not know that $\Delta U = 0$, common sense would suggest that the average energy of the air molecules does not change merely because a proportion of them have passed through a valve.

But $P$ is a substantial number and $\Delta V$ is also a real increment because the air expands as the tyre is let down. Therefore applying equation (2) we have 0 = a substantial number, which is clearly false.

This error (which has been made by Rowan-Robinson (1981), Harwit (1973, 1998) and many others) has been dealt with here in some detail because it is necessary to establish an important principle:

> *Thermodynamics is not like other physical sciences; just because an equation exists does not mean that it is always true. There are some equations which are sometimes true, and sometimes false; their veracity needs to be established by means of a practical experiment or a "thought experiment."*

That is the reason why I introduced the experiment of letting air out of a car tyre.

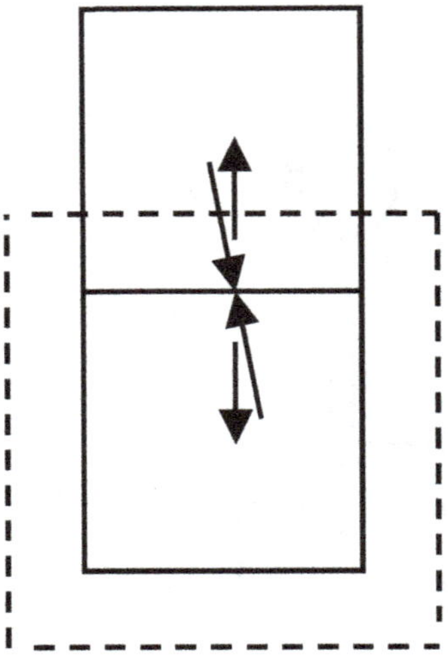

**Figure 2**. A matrix of expanding cubes in the Universe A molecule or atom rebounds off another molecule or atom at the boundary between two cubes.

Cosmologists make the error of applying equation (2) to the expansion of the Universe in this way. The Universe is a self-contained system in which no heat goes in or out of it.

Therefore $Q = 0$. Then equation (2) becomes

$$0 = \Delta U + P\Delta V, \tag{3}$$

or

$$P\Delta V = -\Delta U, \tag{4}$$

*i.e.*, as the volume increases, the internal energy decreases, *i.e.*, the Universe must cool down.

The error in this reasoning is that Eqn. (2) is not a general statement of an adiabatic expansion, it is a special case because in converting $W$ to $P\Delta V$ we are specifying some particular things about our system.

(1) That the substance (usually a gas) is expanding reversibly.
(2) That the work is removed from the system by a machine.
(3) That the machine is 100% efficient thermodynamically.

These features are exemplified in Figure 1.

Equation (2) cannot be applied to the Universe for two reasons. First there is no machine outside the Universe to take away the work. Second, if the Universe expanded in the way postulated by cosmologists, such an expansion would certainly not have been reversible. Actually reversible operations virtually never happen in nature and equation (2) is certainly not applicable to an explosion. It follows that this equation is not applicable to the

"Big Bang" and cosmologists are in error in postulating the Hot Big Bang Theory in this way.

There are a number of other reasons put forward by cosmologists why the expanding Universe should have cooled down. They present the picture of a primeval Universe in which the atoms are moving away from each other—or space between the atoms is expanding; "surely if the relative velocities of atoms are decreasing, this is equivalent to fall in temperature."

I present here a version of this argument which was put to me by Professor P.J.E. Peebles, and then I shall give the refutation which proves that the Universe would not cool down on expansion.

## The proposition: that the universe cools down on expansion

Peebles suggests that the expanding Universe can be considered as an array of cubes, all of which are expanding. In effect there are a number of imaginary partitions which are moving away from each other. These partitions have equal pressures on each side of them. This model is fictitious but legitimate because we can imagine a molecule on one side of the partition bouncing off another molecule on the other side, the impact taking place at the partition. If all molecules approaching the partition are met by equivalent molecules from the other side, this is the same as saying that there is the same pressure on both sides of the partition and this is consistent with our model of the Universe which is expanding but which has uniform pressure at any point in time. (See Figure 2.)

Now as the Universe is expanding, these cubes are expanding, that is to say the opposite faces of these cubes are retreating from each other. A molecule hits a partition, which is retreating, and the molecule bounces back into that cube so that the molecule has a reduced momentum and reduced kinetic energy. This is equivalent to saying that the gas in the cube cools down because molecules are perpetually exchanging kinetic energy by impacts; if a molecule at the boundary of the cube loses some energy this will be reflected in the gas as a whole.

The model is analogous to a gas in a cylinder with a piston (see Figure 1). As the piston retreats, molecules which hit the piston rebound into the gas with reduced momentum and the temperature decreases. This is established experimentally; therefore it is established that the Universe would cool down on expansion.

## The refutation

Let us start from what we know, namely the cylinder and piston (Figure 1). A gas in an expanding cylinder does cool down because molecules hitting the piston lose some of their momentum to the retreating piston. The piston then transfers its momentum and energy to the crankshaft and then to a working machine. Indeed the internal energy of the gas is transferred to the machine quantitatively. Until 40 years ago air separation plants used to achieve their refrigeration by means of reciprocating expansion engines like this and the cooling of the air was achieved measurably by the output of work at the machine, *i.e.* there was a measurable energy balance between the heat energy lost from the air and the work gained at the machine.

Now let us return to our imaginary expanding cubes in the Universe (Figure 2). A molecule hits a retreating partition and rebounds with reduced momentum; but does it transfer some of that momentum to a piston? No! it is simply transferring some momentum to another molecule, which is still in the Universe.

The analogy—that an expanding Universe can be compared with an expanding cylinder and piston—is demonstrably false. The expanding Universe does not lose energy and therefore it does not cool down. This is consistent with known experiments; when an ideal gas expands it does not cool down; the internal energy of an ideal gas is independent of its volume.

This refutation can be summarized this way: the Hot Big Bang Theory does not satisfy an energy balance. This will be dealt with more extensively later where it will be shown that other factors, such as gravitational energy, do not assist the arguments in favour of the Big Bang Theory.

## A summary of errors by cosmologists

Because the Universe is expanding (or at least it appears to be expanding) cosmologists, by applying some fallacious reasoning, arrive at the conclusion that the primeval Universe must have been very hot:

(1) They mis-apply equations relating to the Laws of Thermodynamics.

(2) They state that when a gas expands from high pressure to low pressure, it cools down, which is not necessarily true.

(3) They put forward the concept that the primeval Universe cooled isentropically although this is contrary to the Second Law of Thermodynamics because a spontaneous expansion is accompanied by an increase in entropy.

(4) Cosmologists do not take into account the real physical properties of matter. With the knowledge that the Universe is mostly hydrogen, it is essential that the physical properties of hydrogen are taken into account when cosmologists do calculations on their theories; in fact, very few cosmologists allow for energy changes and entropy changes and volume changes associated with the dissociation and ionization of hydrogen. Later in this paper, rigorous thermodynamic calculations will show that these energy changes and volume changes have a pronounced effect on the supposed cooling of the Universe and that the Hot Big Bang Theory does not conform to the First Law of Thermodynamics. (The point is demonstrated in Figure 4; the graph of internal energy *versus* temperature has several kinks in it; the supposed cooling of the Universe would not have been a steady process).

## The temperature-entropy diagram for hydrogen

What is required is a meaningful method of carrying out thermodynamic calculations and this requires a Temperature-Entropy Diagram for the matter in the Universe, and since the Universe is believed to be 92% hydrogen (atomic percentage) we need a Temperature-Entropy Diagram for hydrogen over the range of astrophysical applications. The point is that to do reliable calculations we shall need data on hydrogen for the following parameters:

- Temperature
- Pressure
- Density
- Enthalpy
- Internal Energy
- Heat of Dissociation
- Heat of Ionization
- Entropy

**Figure 3**
**Temperature-Entropy Diagram for Hydrogen**

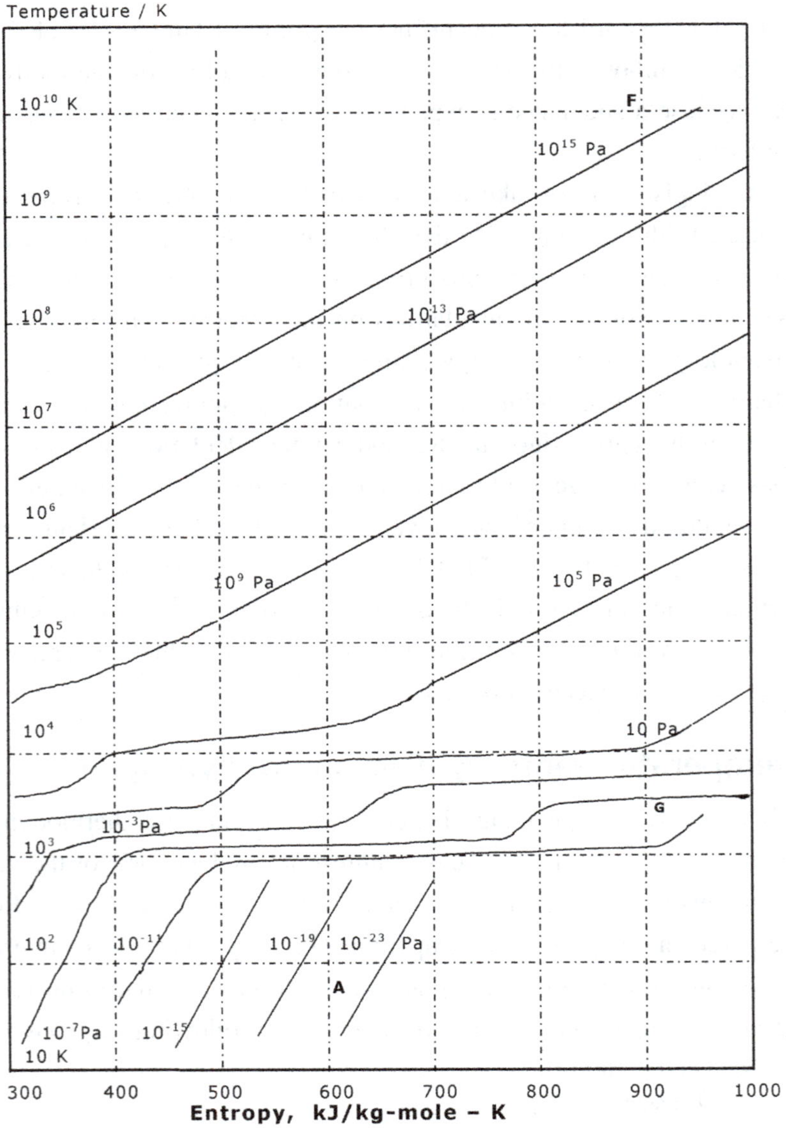


The Temperature-Entropy Diagram is a valuable instrument for studying thermodynamic processes for a number of reasons.

1. In graphical form it presents numerical values for all thermodynamic and physical properties of a fluid (except viscosity and thermal conductivity). It is for the most part based on actual experimental results.
2. Actual operations such as heating, cooling, expansion or compression of a fluid can be traced out clearly and unambiguously. The dia-

gram can then be used for calculating numerical values for energy changes and entropy changes.

3. Most people can appreciate a diagram more easily than the application of equations because of the visual aspects of the diagram. Furthermore many theoretical equations do not portray the real physical properties of a fluid accurately. For example there are no simple equations which relate entropy, enthalpy pressure and temperature to the dissociation and ionization of hydrogen.
4. It becomes evident from the Temperature-Entropy Diagram that certain operations are not permissible, *e.g.*, actions where entropy spontaneously decreases, and this critical approach is not always possible from the application of equations.

There is not space in this paper to display the complete Temperature-Entropy Diagram, a skeleton diagram is given in Figure 3. The complete Temperature-Entropy Diagram for hydrogen is given in the book (Bligh, 2000). The diagram has an extensive range of isobars, lines of constant enthalpy and lines of constant volume. It is based on data provided by the US National Bureau of Standards (1948, 1955, 1961) plus some other experimental results (Kroepelin, 1971) plus some data from Vadya (1960). In order to maintain consistency, we note that a kg-mole of hydrogen has the state

$$H_2 \rightarrow 2H \rightarrow 2H^+ + 2e^-.$$

The diagram is not accurate above $10^6$ K because of relativistic effects, but the general shape is correct. the justification for extrapolating to these high temperatures is that cosmologists such as Alpher (1967), Sciama (1971), P.C.W. Davies (1974) and others do thermodynamic calculations on the primeval hot Universe in which they assume that matter behaves like an ideal gas; in effect my extrapolation makes the same assumption.

## Thermodynamic calculations

This section deals with this subject methodically with detailed calculations and diagrams.

The approach in this chapter is to assume that the Hot Big Bang Theory is true and to use data supplied by its proponents; it is then shown that their data and their reasoning are contrary to the Laws of Thermodynamics. It is explained that the Temperature-Entropy Diagram for hydrogen is a powerful tool both for doing thermodynamic calculations and for analysing operations such as the supposed expansion of the primeval Universe. In particular, it is shown that cosmologists do not use the correct thermodynamic properties of hydrogen in their calculations because they ignore the energy changes which take place when ionized plasma forms hydrogen atoms and when hydrogen atoms associate to form molecules.

A series of calculations on the primeval Universe are presented in which temperature, pressure, internal energy, radiation energy and the expansion scale factor are related.

It is necessary to start with some basic criteria.

(1) The present average density of the Universe; Cosmologists give a range of estimates: Alpher and Herman (1975): $3.41 \times 10^{-31}$ g/cm$^3$; Peebles and Dicke (1968) give a range $1.8 \times 10^{-29}$ g/cm$^3$ to $4.5 \times 10^{-31}$ g/cm$^3$. Sciama (1971) gives a range, middle value $2 \times 10^{-30}$ g/cm$^3$. I take a middle value of $2\times10^{-30}$ g/cm$^3$ = $2\times10^{-27}$ kg/m$^3$. With the approximation that all the matter is molecular hydrogen, this leads to the volume of a kg-mole as $1 \times 10^{27}$ m$^3$.

(2) The temperature of the Universe is extremely variable. The temperatures in the middle of most stars are in the range of $10^6$ to $4 \times 10^7$ K; the temperature of a gas cloud in the Orion constellation has been measured as a little below 80 K (Werner and Harwit); a few very rarefied gas clouds have been found to have temperatures in the order of $10^5$ K; but the consensus among cosmologists is that much of the matter of the Universe is in inter-stellar and inter-galactic regions and is at 10 to 20 K.

(3) I take a notional mean temperature of the Universe as 50 K.
Use the ideal gas equation, $PV = RT$

$$P \times 10^{27} = 8314 \times 50$$
$$P = 4 \times 10^{-22} \text{ Pa.}$$

This calculation provides a point on the Temperature-Entropy Diagram at point A; the idea is that if the Universe had expanded uniformly its present thermodynamic state would be represented by A. The state of matter when stars and galaxies first started to form would not be far removed from A.

(It is accepted that some of the numbers used in this calculation may be in error—perhaps by as much as a factor of 10—but it will be seen at the end of this section that such an error will make no difference to our conclusions. In any case, in astrophysical calculations, the use of rough approximations is commonplace).

(4) We apply the "Peebles model" of the Universe, in which he considers the Universe as a matrix of expanding cubes (Figure 2). Using the data above we calculate that in the present Universe (if it was smoothed out) one kg-mole of hydrogen would occupy a cube of side $10^9$m.

(5) Another key point in our calculations relates to the state of the Universe (according to widely accepted theory) when it changed from being opaque to transparent, which was when the hydrogen changed from being ionic to being atomic. It can be calculated that this was close to 3500 K. Now it is argued that the present background radiation, which is equivalent to 2.7 K, is a relic of this plasma state at 3500 K.

Wien's displacement law gives

$$\frac{\text{temperature of 3500 K}}{\text{temperature of 2.7 K}} = \frac{\text{peak radiation wavelength}}{\text{peak radiation wavelength}} = 1296,$$

*i.e.* there has been a Doppler extension of the wave length by a factor of 1296 which is a measure of linear expansion. Since we have calculated that our "cube containing one kg-mole" now has a side of $10^9$ m, then it had a side of $7.7 \times 10^5$ m at 3500 K.

We use the ideal gas equation, $PV = nRT$, where $n$ represents 2 atoms of hydrogen for one molecule, (because when the Universe became transparent the hydrogen was in the atomic state),

$$P \times (7.7 \times 10^5)^3 = 2 \times 8314 \times 3500$$
$$P = 1.3 \times 10^{-10} \text{ Pa.}$$

This is at the point G on the Temperature-Entropy Diagram.

(6) We continue with our assumption that the Hot Big Bang Theory is correct and trace back the history of a kg-mole of hydrogen to its primeval state, as postulated by Alpher and Herman (1975). They state that at time = 1 second the temperature of the primeval Universe was $1.52 \times 10^{10}$ K and the density of matter was 0.0687 g/cm$^3$, which is equivalent to 34.1 kg-mole/m$^3$.

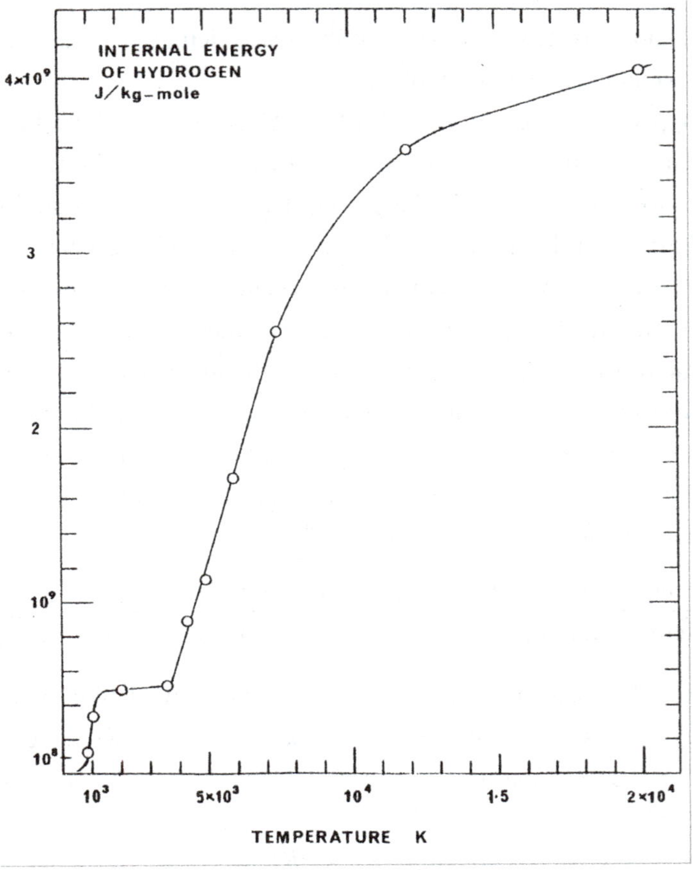


**Figure 4**

But at a temperature of $1.52 \times 10^{10}$ K the hydrogen is ionized to $2H^{+} + 2e^{-}$ (*i.e.* 4 species). Using the ideal gas equation, $PV = nRT$,

$$P \div 34.1 = 4 \times 8314 \times 1.52 \times 10^{10}$$
$$P = 1.72 \times 10^{16} \text{ Pa.}$$

This is specified as point F on the Temperature-Entropy Diagram. This kg-mole would have occupied 0.029 $m^3$ (*i.e.* a cube with a side of 0.31 m). Barrow and Silk (1984, *i*) give similar values for the primeval Universe.

The question now arises "How did that kg-mole pass from its primeval state to its present state?"

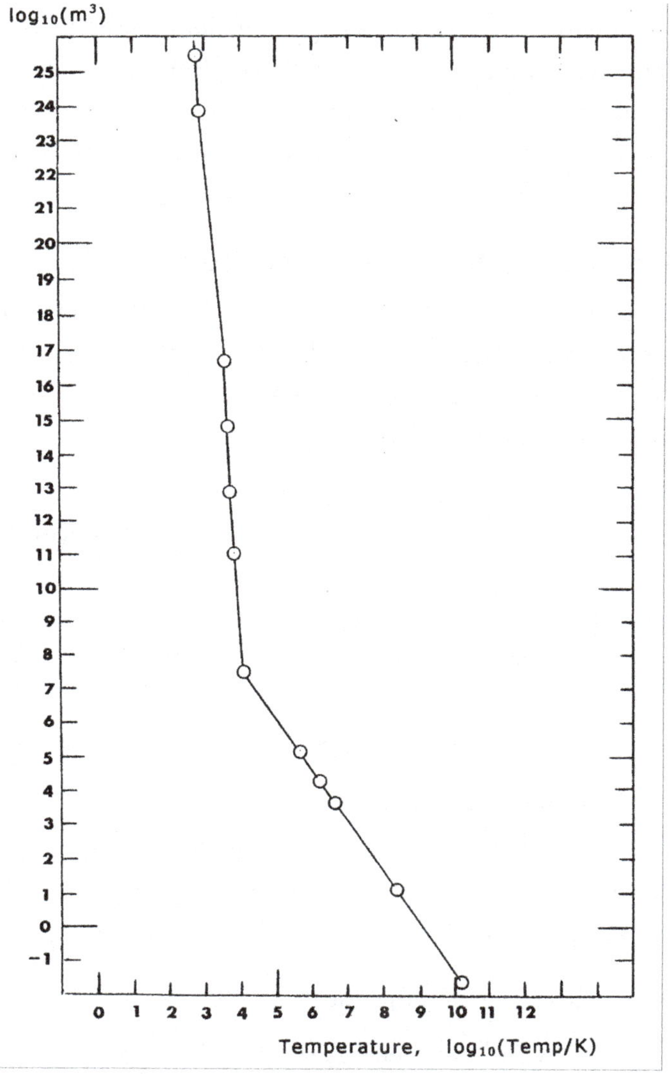


**Figure 5.** This diagram is a plot of volume of one kg-mole of hydrogen versus temperature according to the Hot Big Bang Theory. Note the kink in the curve at about 10,000 K due to the change from ionized hydrogen to atomic hydrogen; this has a surprising effect on radiation energy, see Figure 6.

It is noted that the points F and G are both near a line of constant entropy of about 900 kJ kg-mole$^{-1}$ K$^{-1}$, so it appears that the matter in the Universe passed along a state of constant entropy from $1.52 \times 10^{10}$ to 3500 K and then it cooled to about 50 K where its entropy is about

600 kJ kg-mole$^{-1}$ K$^{-1}$, (point A); (we are simply following the information provided by the proponents of the Hot Big Bang Theory).

In calculations (not given here) the internal energy of hydrogen is plotted against temperature for this state of constant entropy of 900 kJ kg-mole$^{-1}$ K$^{-1}$ in Figure 4.

## Examples of some rigorous calculations

The next stage in the thermodynamic analysis of the Hot Big Bang Theory is to do some rigorous calculations on the state of the Universe as it passed from is primeval state to the state when galaxies and stars started to form.

A typical calculation is presented.

(*i*) Select a value for pressure, *e.g.* $10^5$ Pa.

(*ii*) At a value of entropy = 900 kJ kg-mole$^{-1}$ K$^{-1}$ interpolate the temperature, $4.5 \times 10^5$ K.

(*iii*) Note that the hydrogen is completely ionized, use the ideal gas equation, $PV = nRT$, to calculate the volume of one kg-mole

$$10^5 \times V = 4 \times 8314 \times 4.5 \times 10^5$$
$$V = 1.50 \times 10^5 \text{ m}^3 \text{ kg-mole}^{-1}.$$

This point is plotted on Figure 5.

(*iv*) Calculate the enthalpy, $4.05 \times 10^{10}$ J kg-mole$^{-1}$.

(*v*) Calculate the internal energy, $U = H - PV$

$$U = 4.05 \times 10^{10} - 1.50 \times 10^{10} = 2.55 \times 10^{10} \text{ J kg-mole}^{-1}$$

(*vi*) Calculate the energy of radiation,

$$7.565 \times 10^{-16} \times T^4, \text{ energy} = 3.10 \times 10^7 \text{ J/m}^3.$$

(*vii*) Calculate the energy of radiation relating to one kg-mole of hydrogen, *i.e.* energy

$$\text{J/m}^3 \times V \text{ m}^3\text{/kg-mole} = 4.65 \times 10^{12} \text{ J kg-mole}^{-1}.$$

(This refers to the energy of radiation in space which is occupied by one kg-mole of matter; since the quantity of matter is constant in the epoch under consideration, this is a rational way of computing and presenting the energy of radiation.)

# Table of Results of Thermodynamic Calculations on the Hot Big Bang Theory

The text describes rigorous calculations on hydrogen in the expanding Universe and the basis of these calculations was the physical data provided by Alpher and others. If the Hot Big Bang Theory were true then the Universe would have passed through a temperature range in the first column. The results of this table are plotted in Figures 4, 5 and 6, which show that the Hot Big Bang Theory violates the Law of Conservation of Energy.

| Temperature | Pressure | Volume | Dissociation | Ionization | PV | Enthalpy H | Internal Energy | Radiation Energy | Radiation Energy | Scale Factor |
|---|---|---|---|---|---|---|---|---|---|---|
| K | Pa | $m^3$/kg-mole | $H_2 \rightarrow 2H$ | $H \rightarrow H^+ + e^-$ | Joule/kg-mole | Joule/kg-mole | Joule/kg-mole | Joule/$m^3$ | Joule/kg-mole | $V^{1/3}$/m |
| $1.52 \times 10^{10}$ | $1.7 \times 10^{16}$ | 0.0293 | 1 | 1 | $5 \times 10^{14}$ | $1.26 \times 10^{15}$ | $7.6 \times 10^{14}$ | $4.04 \times 10^{25}$ | $1.18 \times 10^{24}$ | 0.308 |
| $2.256 \times 10^{8}$ | $4.94 \times 10^{11}$ | 15.25 | 1 | 1 | $7.53 \times 10^{12}$ | $1.88 \times 10^{13}$ | $1.13 \times 10^{13}$ | $1.99 \times 10^{18}$ | $3.04 \times 10^{19}$ | 2.48 |
| $3.99 \times 10^{6}$ | $2.43 \times 10^{7}$ | $5.46 \times 10^{3}$ | 1 | 1 | $1.33 \times 10^{11}$ | $3.34 \times 10^{11}$ | $2.01 \times 10^{11}$ | $1.19 \times 10^{11}$ | $1.05 \times 10^{15}$ | 17.6 |
| $1.55 \times 10^{6}$ | $2.19 \times 10^{6}$ | $2.35 \times 10^{4}$ | 1 | 1 | $5.15 \times 10^{10}$ | $1.32 \times 10^{11}$ | $8.035 \times 10^{10}$ | $4.37 \times 10^{9}$ | $1.03 \times 10^{14}$ | 28.7 |
| $4.5 \times 10^{5}$ | $10^{5}$ | $1.50 \times 10^{5}$ | 1 | 1 | $1.50 \times 10^{10}$ | $4.05 \times 10^{10}$ | $2.55 \times 10^{10}$ | $3.10 \times 10^{7}$ | $4.65 \times 10^{12}$ | 53.1 |
| 11990 | 10 | $3.98 \times 10^{7}$ | 1 | 0.996 | $3.98 \times 10^{8}$ | $3.98 \times 10^{9}$ | $3.58 \times 10^{9}$ | 15.6 | $6.22 \times 10^{8}$ | 341 |
| 5922 | $10^{-3}$ | $1.40 \times 10^{11}$ | 1 | 0.420 | $1.40 \times 10^{8}$ | $1.94 \times 10^{9}$ | $1.80 \times 10^{9}$ | 0.930 | $1.30 \times 10^{11}$ | $5.19 \times 10^{3}$ |
| 4905 | $10^{-5}$ | $9.86 \times 10^{12}$ | 1 | 0.209 | $9.86 \times 10^{7}$ | $1.23 \times 10^{9}$ | $1.13 \times 10^{9}$ | 0.438 | $4.32 \times 10^{12}$ | $2.14 \times 10^{4}$ |
| 4324 | $10^{-7}$ | $7.97 \times 10^{14}$ | 1 | 0.132 | $7.97 \times 10^{7}$ | $9.78 \times 10^{8}$ | $8.98 \times 10^{8}$ | 0.243 | $1.94 \times 10^{14}$ | $9.27 \times 10^{4}$ |
| 3540 | $10^{-9}$ | $6.07 \times 10^{16}$ | 1 | 0.03 | $6.07 \times 10^{7}$ | $5.80 \times 10^{8}$ | $5.19 \times 10^{8}$ | 0.119 | $7.21 \times 10^{15}$ | $3.93 \times 10^{5}$ |
| $T_m$ 820 | $10^{-17}$ | $8.68 \times 10^{23}$ | 0.273 | 0 | $8.68 \times 10^{6}$ | $1.45 \times 10^{8}$ | $1.36 \times 10^{8}$ | | | $9.54 \times 10^{7}$ |
| $T_r$ 28.3 | | | | | | | | $4.85 \times 10^{-10}$ | $4.21 \times 10^{14}$ | |
| $T_m$ 50 | $4.2 \times 10^{-22}$ | $1 \times 10^{27}$ | 0 | 0 | $4.2 \times 10^{5}$ | $2.1 \times 10^{6}$ | $1.68 \times 10^{6}$ | | | $1 \times 10^{9}$ |
| $T_r$ 2.7 | | | | | | | | $4.02 \times 10^{-14}$ | $4.02 \times 10^{13}$ | |

$T_m$ = temperature of matter.
$T_r$ = temperature of radiation.

These two temperatures are different only when the system is below 3500 K and it is "un-coupled."

(*viii*) The linear expansion of the Universe is proportional to the cube root of the volume of a kg-mole, $V^{1/3} = 53.1$ m kg-mole$^{-1}$.

The results of a series of such calculations are given in the Table and are plotted in Figures 5 and 6.

In Figure 5 the volume of one kg-mole is plotted against temperature. The important feature to note is that the volume does not increase uniformly as the temperature falls, there is a kink at $10^4$ K. This has a surprising effect.

The energy of radiation is proportional to $T^4 \times$ Volume.

Peebles' model of the Universe is applied, in which he considers the Universe as a matrix of expanding cubes. Each cube contains one kg-mole of hydrogen (or the precursor of hydrogen) and therefore the cube root of the volume is a scale factor of the linear expansion of the Universe.

Therefore the energy of radiation is plotted against $V^{1/3}$ in Figure 6 (the scales are logarithmic) and the curve shows a marked change in direction at $T = 10^4$ K. There is a steep fall in radiation energy followed by a steep increase. The graph for internal energy of the hydrogen also shows an inflexion. The reason for the inflexions in the curves is the change in properties of the hydrogen due to dissociation and ionization.

Below 3500 K when hydrogen is not ionized the Universe is said to be "de-coupled" and the radiation temperature, $T_r$, is no longer the same as the temperature of matter, $T_m$. Therefore Wien's Law has been used to calculate the radiation temperature and the radiation energy for a calculated volume of a kilogram-mole of hydrogen. These values are recorded in the Table and in Figure 6.

In its present form the theory of the expansion of the Universe cannot possibly be true because it is contrary to the Law of Conservation of Energy (note that the energy associated with radiation falls by 15 orders of magnitude and the internal energy of matter falls by 8 orders of magnitude and these enormous discrepancies far outweigh any errors in estimating the present density of the Universe).

Basu and Lynden-Bell (1990) attempt to explain the relatively low entropy of the present Universe compared with the high entropy of the primeval Universe by means of the generation of radiation with a high entropy; but their paper does not attempt an energy balance and therefore they did not identify the violation of the First Law of Thermodynamics.

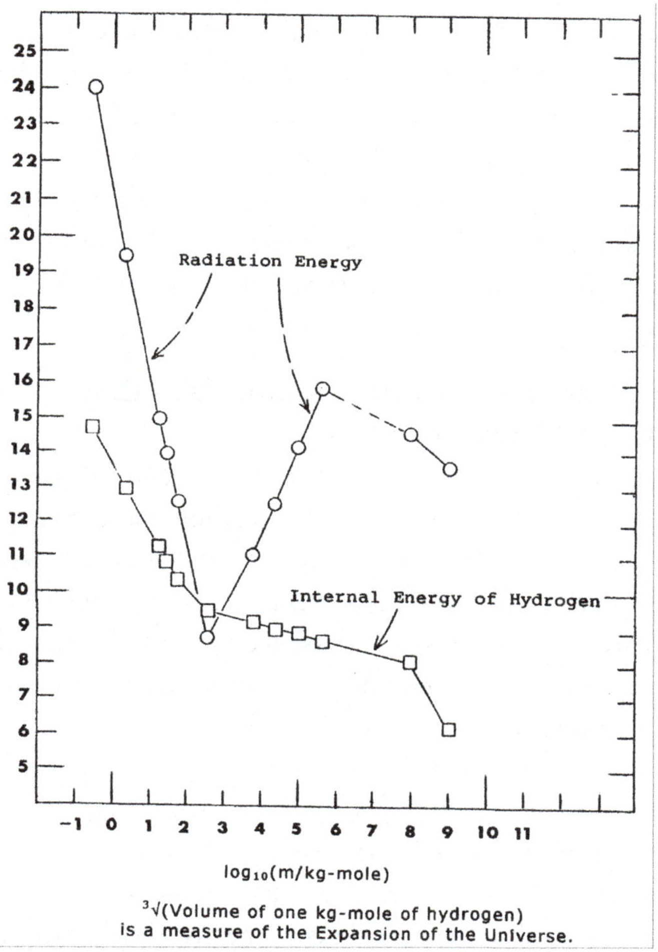


**Figure 6.** Energy in Joules (log scale) associated with one kg-mole of hydrogen is plotted against the Expansion of the Universe. These curves show there is no conservation of energy.

Now it is sometimes argued that the thermal energy of matter in the expanding Universe decreases because the gravitational potential energy increases (P.C.W. Davies, 1974, and Goldberg and Scadron, 1981), but such a theory cannot possibly explain the alternate fall and rise in radiation energy (Figure 6). Up till now this feature has not been noticed because cosmologists have not done these rigorous thermodynamic calculations.

There are other objections to the theory that the internal energy falls because gravitational energy increases. The heat given up is proportional to the specific heat capacity of the primeval fluid, whereas the work against gravitation is a function of the gravitational constant and the mass of the Universe. It would be an extraordinary coincidence if these entirely unrelated properties were to balance exactly. Furthermore the heat given up has inflexions due to the energy of ionization and the energy of dissociation (Figure 4) which militate against the gravitational argument even more strongly.

The inevitable conclusion is that the Hot Big Bang Theory cannot possibly be true.

## Theorem; That The Universe Cannot Have Cooled Down By Way of an Isentropic Expansion

It is an essential postulate of the "Hot Big Bang Theory" that the hot primeval universe cooled down by reason of expansion and it has already been shown that many of the reasons given by cosmologists are incorrect. But there is one argument that needs to be examined again, namely that the cooling of the Universe comes about because the Universe is expanding against the force of gravity, *i.e.* the internal energy of matter is decreasing because gravitational potential energy is increasing.

The basis of this critique is two alternative "thought experiments" in which we consider the expanding Universe which contains a cylinder and piston. The cylinder is thermally insulated from the Universe. In "Case A" the Universe expands from time, $t_1$, to time, $t_2$, while some high-pressure fluid remains encapsulated in the cylinder. The piston then expands the fluid isentropically to the pressure of the expanded Universe such that work, $W$, is done. This work is stored in some device such as a metal spring or an electrochemical cell. During the expansion of the piston the Universe has expanded to volume, $V_2$. The diagrams in Figure 7 are purely schematic, they are not intended to imply that the Universe has an edge.

In "Case B" the Universe expands to volume, $V_2$, and the fluid in the cylinder expands at the same time. No work is done and no energy is stored.

If the Universe expands isentropically, then the fluid in Case B (both inside and outside the cylinder) is in the same state as Case A (both inside and outside the cylinder). The expansion of the fluid inside the cylinder of Case A is undoubtedly isentropic (by definition). But Case A has gained some work, $W$, which Case B does not have. Moreover, the work, $W$, could be used to recompress the fluid in the cylinder which could not be achieved in Case

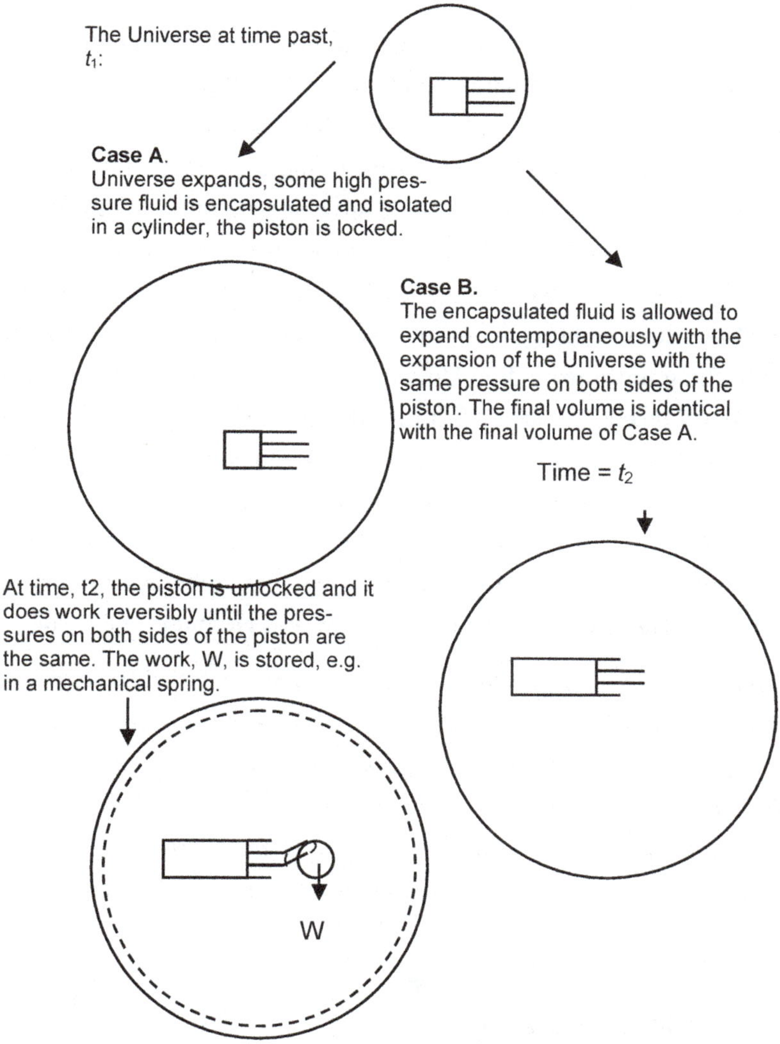


**Figure 7**. The Expanding Universe

B. Clearly the two cases are different; therefore by *reductio ad absurdum* the expansion of the Universe cannot be isentropic.

As has already been mentioned, it is argued by some cosmologists (*e.g.*, P.C.W. Davies, 1974, and H.S. Goldberg and M.D. Scadron, 1981) that the expansion of the Universe is isentropic because it is doing work against gravity. Other cosmologists argue that the matter in the universe is expanding isentropically because it is counter-balanced by the increase in entropy caused by the generation of radiation into space (R. Penrose and Basu and Lynden-Bell). But in the two Cases, A and B, the final volume is the same for both. Therefore, the gravitational state and the radiation state are the same for both. Therefore these alternative explanations for the cooling of the Universe do not explain the paradox that Case A has gained some energy which Case B does not have. Since the expansion in the cylinder is certainly isentropic it follows that the expansion of the Universe cannot be isentropic.

In reality the expansion of the Universe (whether against gravity or not) is spontaneous and not isentropic. The real difference between these two cases is that the fluid in the cylinder in Case A does extra work and therefore it cools more than the surrounding Universe. Hence the extra work in Case A is gained from the sensible heat of the fluid. The paradox is resolved by the well-established thermodynamics that the spontaneous expansion is not isentropic and therefore the Universe would not cool in this way.

I have studied scores of scientific papers and books on Astrophysics and Cosmology and I have found only one paper which recognized this problem with the Hot Big Bang Theory. Alpher, Gamow and Herman (1967) state

> *It may be noted that for this cosmological model in which matter is conserved, the total energy is not; work done in the adiabatic expansion is readily calculable but not readily accountable in terms of whence it goes, since the calculation proceeds as though the matter-radiation mix were doing work on a container in the expansion—a container of somewhat dubious reality.*

## Summary and Conclusions

This paper presents the reasons why the Hot Big Bang Theory of the origin of the Universe is thermodynamically unsound. With the aid of a Temperature-Entropy Diagram for hydrogen, thermodynamic calculations are done on the Hot Big Bang Theory which have not been done previously by cosmologists.

Those cosmologists who still believe in the Hot Big Bang Theory need to explain the following conundrums.

(1) Justify the Peebles model of the Universe in which he suggests that a matrix of expanding cubes would explain a fall in temperature; my critics must disprove my "Refutation."

(2) The application of the Temperature-Entropy Diagram for hydrogen shows that the entropy of the Universe has decreased very substantially according to the Hot Big Bang Theory. This is contrary to the Second Law of Thermodynamics because this law states that entropy must increase for a spontaneous expansion. Cosmologists must explain this violation of the Second Law of Thermodynamics.

(3) Some cosmologists (but not all cosmologists) justify the cooling of the Universe in terms of exchange between internal energy and gravitational potential energy. I point out that this theory requires a close numerical balance between two completely unrelated properties, the specific heat capacity of the matter in the Universe and the gravitational constant. Furthermore, Figure 4 shows that there is not a smooth curve between internal energy and temperature. This point is proved with even more force in Figures 5 and 6. Cosmologists need to show how gravitational energy can follow these inflexions.

(4) The alleged high temperature of the primeval Universe also contained enormous radiation energy. With the expansion of the Universe, this radiation energy must have gone through a steep fall, then a steep rise followed by another fall, Figure 6. This is clearly contrary to the First Law of Thermodynamics (the law of conservation of energy). Cosmologists need to explain this paradox.

(5) Cosmologists need to find a fallacy in my theorem that "A Hot Primeval Universe Would Not Cool Down Isentropically On Expansion" (Figure 7).

## References


Alpher, R.A., Gamow, G. and Herman, R. (1967) "Thermal Cosmic Radiation and the Formation of Protogalaxies," *Proc. of the National Acad. of Sci.* Vol. 58, pp. 2179 – 2186.

Alpher, R.A. and Herman, R. (1975) "Big Bang Cosmology and Cosmic Black-Body Radiation," *Proc. of the Am. Phil. Soc.* Vol. 119, No. 5, pp. 325 – 348.

Barrow, J.D. and Silk, J. *The Left Hand of Creation* (W. Heinemann, London 1984) (i) p. 85, estimate of density of primeval Universe, (ii) p.99 states that the early Universe contained helium 25% by mass.

Basu, B. and Lynden-Bell, D. (1990), "A Survey of Entropy in the Universe" *Quarterly Journal of the Royal Astronomical Society*, Vol. 31, p.359.

Bligh, B.R. (2000) *The Big Bang Exploded! Cosmology Corrected, A Commentary with Thermodynamics*; this book contains an extensive Temperature-Entropy Diagram for hydrogen which demonstrates the thermodynamic properties of hydrogen, including the lines of constant enthalpy. Large scale copies of the diagram can be obtained from the author, e-mail brbligh@hotmail.com.

Davies, P.C.W. (1974), *The Physics of Time Asymmetry* (Surrey University Press), p. 82 and p. 109.

Goldberg, H.S. and Scadron, M.D. *Physics of Stellar Evolution and Cosmology* (Gordon and Breach Science Publishers, 1981) p. 326.

Kroepelin, H. *et al. Thermodynamic Diagrams for High-Temperature Plasmas of Air, Air-Carbon, Carbon-Hydrogen Mixtures and Argon* (Braunschweig, 1971).

NBS, Woolley, H.W., Scott, R.B. and Brickwedde, F.G., (1948, Nov) "Compilation of Thermal Properties of Hydrogen and its Various Isotopic and ortho-para Modifications." *J. Research National Bureau Standards*, Vol. 41, pp. 379 – 475.

NBS Circular 564, (Nov. 1955), Hilsenrath, J. *et alia*, *Tables of Thermal Properties of Gases.*

NBS Technical Note 120, (Nov. 1961), Dean, J.W. *A Tabulation of Thermodynamic Properties of Hydrogen for Low Temperature to 300 K and from 1 to 100 atm.*

Peebles, P.J.E. "Recombination of the Primeval Plasma," *Astrophysical Journal*, Vol. 153, pp. 1 – 11, July 1968.

Peebles and Dicke, "Origin of the Globular Star Clusters," *Astrophysical Journal*, Vol. 154, p. 891, December 1968.

Penrose, R. *The Emperor's New Mind*, p. 447, (Publ. Vintage 1990).

Rowan-Robinson, M. *Cosmology* 2nd Ed. 1981, 3rd Ed. 1996, (Oxford Clarendon).

Sciama, D.W. *Modern Cosmology* (1971, Cambridge University Press).

Vardya, M.S., (1960, March) "Hydrogen-Helium Adiabats for Late-Type Stars," *Astrophys. J. Supplement*, Vol. 4, pp. 281 – 336.

Werner, M.W. and Harwit, M. "Observational Evidence for the Existence of Dense Clouds and Inter-stellar Molecular Hydrogen." *Astrophysical Journal*, Vol. 154, p. 881, December 1968 (note, this paper refers to hydrogen gas clouds near the Orion nebula).

# On Possible Experimental Evidence for a Breakdown of Local Lorentz Invariance[16]

Fabio Cardone[17] and Roberto Mignani[18]

We report the preliminary results of an experiment aimed at detecting a DC voltage across a conductor induced by the steady magnetic field of a coil. Two experimental runs, carried out with different apparati, showed positive evidence for such an effect, which might be interpreted as a breakdown of local Lorentz invariance. The new limits obtained by this new class of experiments are fully compatible with those already present in the literature for LLI effects.

## 1 - Introduction

The fundamental teaching of Einstein's relativity theories is that physical phenomena occur in four-dimensions (three spatial and one time dimension), space-time possessing a *global curved* (Riemannian) structure and a *local flat* (Minkowskian) structure. However, it is a long-disputed problem whether local Lorentz invariance (LLI) preserves its validity at any length or energy scale (far enough from the Planck scale, when quantum fluctuations are expected to come into play). Doubts as to the reliability of a Lorentz-invariant description of physical phenomena at subnuclear distances were, *e.g.,* put forward in the mid-sixties, even in standard (and well-known) textbooks[(1)].

From the experimental side, the main tests of LLI can be divided into roughly three groups[(2)]:

a) Michelson-Morley (MM)-type experiments, aimed at testing isotropy of the round-trip speed of light;

b) Tests of the isotropy of the one-way speed of light (based on atomic spectroscopy and atomic timekeeping);

---

[16] Invited talk at the Conference on Relativity Theory, Cesena, Italy, Febr. 1999.

[17] Dipartimento di Fisica, Universita' dell'Aquila, Via Vetoio, 67010 Coppito, L'Aquila, Italy, and I.N.D.A.M. - G.N.F.M.

[18] Dipartimento di Fisica "E.Amaldi," Universitá di Roma "Roma Tre," Via della Vasca Navale, 84, 00146 Roma, Italy, and INFN, Sezione di Roma III; e-mail: mignani@fis.uniroma3.it

c) Hughes-Drever-type (HD) experiments, testing the isotropy of nuclear energy levels.

All such experiments set upper limits on the degree of violation of LLI.

From the theoretical side, many generalizations of Special Relativity and/or LLI breaking mechanisms can be found in the literature. A brief account of the main ones can be found in [3]. Very interesting approaches to LLI breakdown within the framework of the Standard Model have been recently considered by Coleman and Glashow(4,5), with the proposal of new tests of Special Relativity in cosmic-ray and neutrino physics, and by Jackiw(6), who puts very stringent limits on such effects.

A number of years ago, we have proposed a generalization of SR based on a "deformation" of space-time, assumed to be endowed with a metric whose coefficients depend on the energy of the process considered(3,7). Such a formalism (*Deformed Special Relativity*, DSR) applies in principle to *all* four interactions (electromagnetic, weak, strong and gravitational) and provides a metric representation of them (at least for the process and in the energy range considered)(3). DSR predicts, among the others, different maximal causal (*i.e.*, maximum attainable) speeds for different interactions and/or different systems (in agreement with the results of refs. [4,5]). Moreover, it was shown that such a formalism is actually a five-dimensional one, in the sense that the deformed Minkowski space is embedded in a larger Riemannian manifold, with energy as the fifth dimension(8).

The important point to be stressed is that the DSR formalism was not introduced on a mere speculative basis, but was motivated by the apparent inadequacy of the standard SR to fully and consistently describe some physical processes. These are: the lifetime of the (weakly decaying) $K^0_S$-meson; the Bose-Einstein correlation in (strong) pion production; superluminal photon tunnelling. All such phenomena apparently admit of a consistent interpretation in terms of deformed, energy-dependent metrics(3). Moreover, an analogous description seems to hold for gravitation, as well(9), on the basis of the experimental results on the slowing down of clocks in a gravitational field(10). *All these results seem to provide a first* (although preliminary), *indirect evidence for a breakdown of local Lorentz invariance for all fundamental interactions.*

Quite recently the present authors, together with U. Bartocci(11), proposed a new electromagnetic experiment aimed at testing LLI and capable of providing *direct evidence* for its breakdown. The results obtained in a preliminary experimental run carried out in June 1998—essentially aimed at providing new upper limits on the LLI breakdown parameter by an entirely new class of electromagnetic experiments—admit as the most natural interpretation the fact that *local Lorentz invariance is in fact broken*(11,12).

The experiment was just repeated some months ago with an improved apparatus. The preliminary analysis of this second run seems to confirm the positive evidence of the previous run(13).

In this paper, we want to review the basic results of these two experimental runs, and discuss them.

The organization of the paper is as follows. In Section 2 we describe the experimental set-up and the results obtained. Section 3 contains the analysis and discussion of these results.

## 2 - Experimental set-up and results

The new proposed test is based on the possibility of detecting a non-zero Lorentz force between the magnetic field **B** generated by a stationary current $I$ circulating in a closed loop $\gamma$, and a charge $q$, on the assumption that both $q$ and $\gamma$ are at rest in the same inertial reference frame. Such a force is zero, according to the standard (relativistic) electrodynamics.

The experimental set-up was devised in order to put new upper limits on the breakdown of LLI, by means of such an entirely new class of electromagnetic experiments, and also to test possible anisotropic effects in such limits.

The experimental device used is schematically depicted in Fig. 1. It consisted of a Helmholtz coil $\gamma$ and a Cu conductor R placed inside it on a plane orthogonal to the $\gamma$ axis. The conductor R was connected in series to a capacitor C, and a voltmeter was connected in parallel to the capacitor, so as to measure the voltage due to a possible gradient of charge across R. The conductor could change its orientation in the coil plane. Moreover, the whole system of the RC circuit and the coil could turn to make its plane coincide with one of the coordinate planes. The centre of the geometrical coordinate system coincided with the centre of the coil. The coordinate system was chosen as follows: the $(x,y)$-plane tangent to the Earth surface, with the $y$-axis directed as the (local) Earth's magnetic field $\mathbf{B}_T$; the $z$-axis directed as the

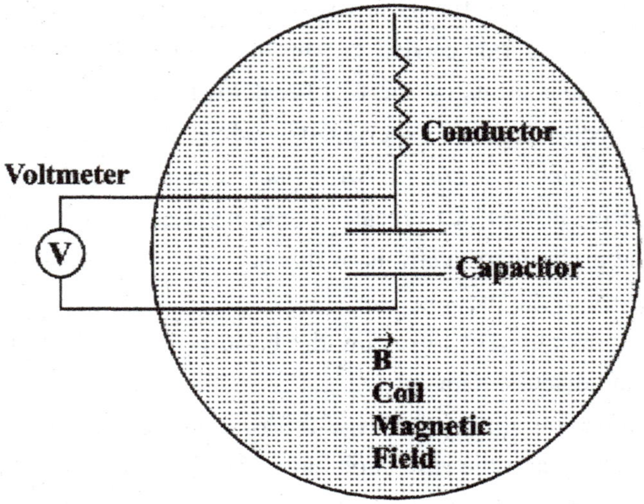


**Fig. 1.** Experimental setup. See the text.

outgoing normal to the Earth's surface, and the $x$-axis directed so that the coordinate system is left-handed. The conductor orientation in the plain coil was parameterized in terms of an angle $\alpha$ (ranging from 0 to $2\pi$). The rotation of $\alpha$ was chosen clockwise in the plain coil with respect to an observer oriented along the coordinate axis orthogonal to the coil plane. The first orientation of R corresponding to the angle $\alpha = 0$ was along the negative direction of the $z$-axis in the case of the two vertical canonical planes (see Fig. 2 for the $(x, z)$-plane), and along the negative direction of the $y$-axis in the case of the horizontal plane. A steady-state current $I$ circulating in the coil produced a constant magnetic field **B** in which the RC circuit is embedded. The circuit and the coil were mutually at rest in the laboratory frame.

Measurements of the voltage $V$ across the capacitor were carried out for the system lying in the different coordinate planes $(x,y)$, $(x,z)$, $(y,z)$, and at different values of the orientation angle $\alpha$ of the circuit in the plane considered (spaced by $\pi/4$). The orientation of the coil $\gamma$ and the direction of the current $I$ were chosen so that, when $\gamma$ lies on $(x,y)$, its magnetic field **B** is directed as $z$; when $\gamma$ is on $(y,z)$, **B** is directed as $x$; for $\gamma$ on $(x,z)$, **B** is directed as $\mathbf{B}_T$. The last arrangement of the apparatus is shown in Fig. 2.

The measurement runs were carried out on three different days (each day with a different orientation of the apparatus plane), two times a day.

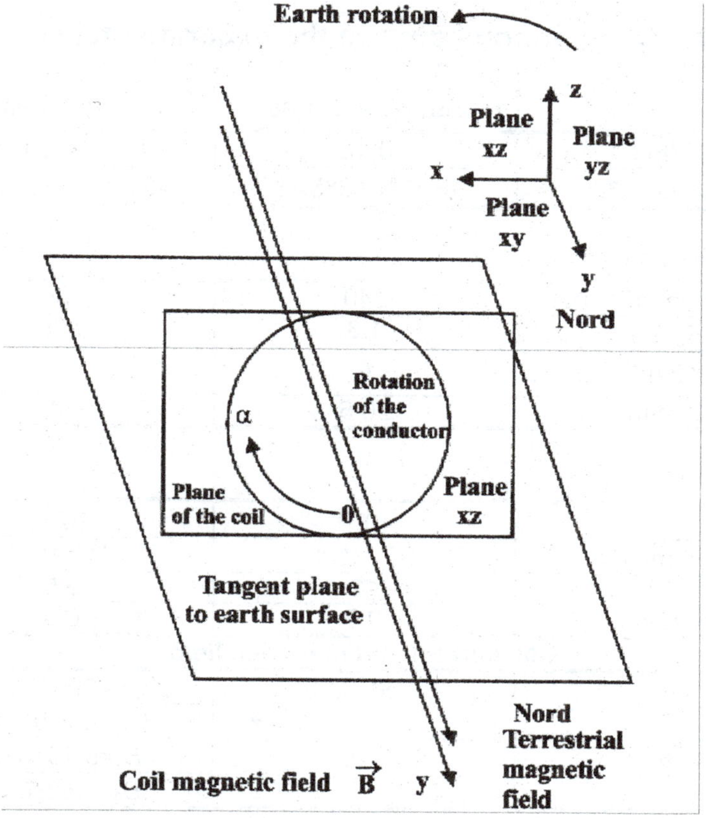


**Fig. 2**. Schematic view of the orientation of the apparatus and the related magnetic fields in the $(x,z)$ plane.

Every run consisted of five measurements of the voltage taken at the same orientation angle $\alpha$, for eight values of $\alpha$ in $(0,2\pi)$. For a fixed angle, the five measurements of $V$ were taken at time intervals of 60 sec from each other.

The number of measurements and their time intervals were previously fixed according to the thermal stability of both the voltmeter and the coil (the temperature in the laboratory was comprised between 24°C and 25°C). The procedure used to determine the optimal number of measurements for each angle $\alpha$ was as follows. The system was probed by means of a known voltage. Then, we determined the mean value of the (poissonian) distribution of the number of measurements able to reproduce the known value of the voltage with a statistical error equal to the experimental measurement error of the known signal. Such a mean value is just the required optimal number of measurements (five). An analogous procedure was followed in order to de-

**Table 1 – Comparison between the experimental setups**

| | First Experiment - 1998 | Second Experiment - 1999 |
|---|---|---|
| $B_L$ (mT) | 0.16 | $B_L$=$B_T$=0.046 |
| coordinates | 46°6'N 13°52'E | 42°30'N 13°56'E |
| **Helmholtz coil** | | |
| $N$ (number of turns) | 240 | 198 |
| $R_c$ (Ω) | 1.8 | 0.66 |
| $\ell$ (cm) | 4 | 4 |
| $D$ (cm) | 26.5 | 14 |
| **Circuit** | | |
| $C$ (Farad) | $10^{-12}$ | $10^{-12}$ |
| $R$ (mΩ) | 36.7±0.1 | 21.1±0.6 |
| $L$ (cm) | 12 | 6.5 |
| $d_R$ (mm) | 0.5 | 0.5 |
| **Coil current and magnetic field** | | |
| $I$ (A) | 5.000±0.001 | $I_1$=4.000±0.001 |
| | | $I_2$=2.000±0.001 |
| $B$ (mT) | 3.65±0.05 | $B_1$=5.14±0.01 |
| | | $B_2$=2.58±0.01 |
| **Multimeter** | | |
| type | Simpson 474 | HP 34401A |
| sensitivity | $10^{-5}$V | $10^{-7}$V |
| **Instrumental zero** | | |
| $V_0$ (μV) | 15.0±10.0 | 16.5±1.5 |

termine the time interval. In particular, a shorter time interval was checked to give too many fluctuations; a longer one was useless, because the instruments already attained thermal stability.

The tests to find the time interval and the number of measurements were carried out (at known voltage) both for $\mathbf{B} = 0$ and for $\mathbf{B} \neq 0$, by checking in the latter case that the actual value of **B** did not affect the behaviour of the instruments used for the measurements (so that the known voltage was determined with the same precision as in the former case).

The zero level of the voltmeter was fixed by means of the same procedure followed for the time interval between measurements and for the number of measurements. The measured values of the voltage $V$ were assumed to represent a physically acceptable, non-zero signal *only if external* to the above interval. Clearly, this made it possible to eliminate (at least most of) the fluctuation contributions and other spurious effects connected with the background.

The second experiment was carried out in 1999 in a different place, with a different apparatus and with a sensitivity improved by two orders of magnitude. Two different values of the coil current $I$ (one halved with respect to the other), and therefore of the coil magnetic field, were considered. The comparison between the two experiments is given in Table 1.

In Table 1, $\mathbf{B}_L$ denotes the local magnetic field. In the first experiment, it was found that $\mathbf{B}_L > \mathbf{B}_T$ due to the unavoidable presence of metallic masses in the laboratory; however, the direction and magnitude of the two magnetic fields coincided ($\mathbf{B}_L/B_L = \mathbf{B}_T/B_T$). In the second experiment, it was found instead that $\mathbf{B}_L = \mathbf{B}_T$.

In the first experiment, the measurements performed with the system lying on the planes $(x,y)$ and $(y,z)$ gave values of $V$ compatible with the instrument zero. Indeed, in such cases the statistical tests of correlation showed that each of the points outside the zero-voltage band is uncorrelated with the preceding and the subsequent point either, and the whole set of points was shown to be uncorrelated ($R^2 < 30\%$). Let us recall that—as stressed before—each point is the average of five measurements, taken at the same angle. As to the measurements in the plane $(x,z)$, it was shown instead that *the four points outside the zero band are statistically correlated* ($R^2 > 80\%$), *and so they represent a valid candidate for a non-zero signal.*

A polynomial interpolating curve for these points is shown in Fig. 3. Such an interpolating procedure was essentially aimed at finding the angle $\alpha_{max}$ corresponding to the maximum value of $V$, $V^{xz}{}_{max} = (3.6 \pm 1.0) \times 10^{-5}$volt. The value found was $\alpha_{max} = 3.757$ rad. The knowledge of $\alpha_{max}$ is needed in order to determine the value of the anisotropic LLI violation parameter in our case[(12)].

In the second experiment, a signal candidate was analogously found in the plane $(x,z)$. For $B = B_I = (5.14 \pm 0.01)$mT. The average peak value was in excellent agreement with the result of the former experiment: $V^{xz}{}_{max} = (3.54 \pm 0.01) \times 10^{-5}$volt. The signal was again highly anisotropic, and its behaviour with $\alpha$ is the same as depicted in Fig. 3. However, *possible*

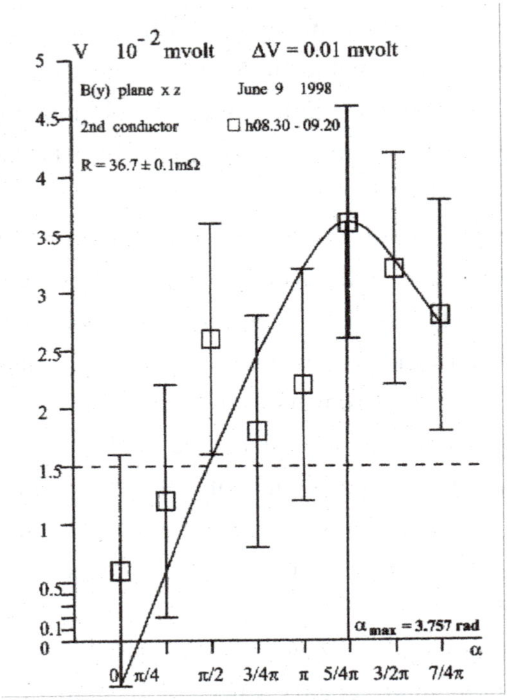


**Fig. 3.** Curve interpolating the data obtained with the apparatus in the $(x,z)$ plane, showing the angle of maximum signal $\alpha_{max}$=3.757rad.

*signal candidates were now also found in the planes (x,y) and (y,z)* (this is also a consequence of the higher sensitivity of the multimeter, improved by two orders of magnitude). In those planes, there was no dependence on $\alpha$, and therefore no spatial anisotropy. On the contrary, a time anisotropy was found in the $(x,y)$ plane, since the measurements taken a.m. gave values within the instrument zero. The average level values found for $B = B_1$ were $V'_{xy} = (3.07 \pm 0.01) \times 10^{-5}$ volt and $V'_{yz} = (2.66 \pm 0.01) \times 10^{-5}$ volt. The measurements taken with the halved value of the coil magnetic field, $B = B_2 = (2.58 \pm 0.01)$ mT, gave similar results, with voltage values $V^{xz}_{max} = (4.18 \pm 0.01) \times 10^{-5}$ volt, $V'_{xy} = (3.44 \pm 0.01) \times 10^{-5}$ volt and $V'^{yz} = (3.06 \pm 0.01) \times 10^{-5}$ volt. Not only were these values not halved with respect to those obtained for $B = B_1$ (as expected in the case of a linear relation between $V$ and $I$, like that derived *via* the Lorentz force), but, surprisingly enough, they were slightly higher! Moreover, a check was made by reversing the coil current. No change in the sign of $V$ occurred. This allows us to conclude that the *effect we observed is independent of the magnitude and direction of the current.*

Precautions taken to avoid false signals included: shielding from external stray fields, provided by the structure of the laboratory building itself; a suitable geometry of wires and connections of the apparatus, able to avoid self-inductance effects; continuous monitoring of the local magnetic field $\mathbf{B}_L$ during the experiments (by measuring it at the beginning and at the end of each measurement run, in order to test its stability as well), that allowed us to rule out effects due to fluctuations of $\mathbf{B}_L$; checking the stability of both the coil magnetic field **B** (including its variations in direction and magnitude), and the current $I$. Effects due to the voltmeter stability with temperature (for $T = 25 \pm 1$°C) were actually negligible. A possible influence of the magnetic field of the Sun and/or the solar wind could also be disregarded. As already stressed, background effects were taken into account in fixing the instrumental zero. The repetition of the experiment in a different place was aimed at getting rid of possible inescapable local effects.

## 3 - Discussion and conclusions

First of all, let us stress that some authors(14-16) have foreseen effects analogous to what we observed. In refs. [14,16], it is shown that a non-zero electric field is expected to exist outside wires and/or closed loops carrying a constant current, whereas a non-null Lorentz force between a charge and a coil both at rest in the same reference frame is predicted in ref. [15] by the classically interpreted Maxwell theory. Moreover, some claims of evidence for such anomalous electromagnetic phenomena are found in the literature(17–19), although they are controversial(20,21). However, all such (both theoretical and experimental) effects *do depend on the magnitude and/or the direction of the current*, hence are fully isotropic, and therefore have nothing to do with our effect.

Among the possible interpretations, we suggest that the effect is due to a kinematical decoupling of the magnetic field ***B*** from the coil that generates it. As a consequence, the coil and the conductor are at rest in the same frame (the laboratory frame), whereas the field ***B*** is at rest with respect to an absolute reference frame $\Sigma_0$. The existence of (and the motivations for) such a frame has been recently revived in the literature(22). Possible candidates for $\Sigma_0$

are: *a*) The frame where the 2.7 K background thermal radiation is isotropic for all the velocities of light; *b*) the Hubble frame, where an observer would see all galaxies receding away with the Hubble expansion velocity; *c*) the frame tied to the moving arm of our Galaxy; *d*) the frame of the stochastic background gravitational radiation[(23)].

In the framework of this interpretation, it is possible to give an estimate of the Earth's speed $v$ with respect to such an absolute frame. We obtain[(11,12)]

$$v = (5.906 \pm 0.001) \times 10^{-2} \text{ m/sec}.$$

It is now easy to see why it is impossible to detect such an effect by means of an experiment of the Michelson-Morley-type. As is well known, the displacement $\Delta n$ of the interference fringe in a MM experiment is given by

$$\Delta n = (\ell_1 + \ell_2)\lambda^{-1}(v_R/c)^2$$

where $\ell_1$, $\ell_2$ are the lengths of the arms of the interferometer, $\lambda$ is the light wavelength, and $v_R \cong 3 \times 10^8$ m/sec is the velocity of Earth's revolution. In the original MM experiment, it is $\ell_1 + \ell_2 = 22$ m, $\lambda = 5.5 \times 10^{-7}$ m, $\Delta n = 0.4$. In our case, we have to replace $v_R$ by the Earth's speed $v$ with respect to the absolute reference frame $\Sigma_0$, whose value, according to our experimental findings (and the interpretation we proposed), is given by the above estimate, $v \cong 0.06$ m/sec ). Then, by using the same parameters of the original MM experiment, one gets

$$\Delta n \cong 0.2 \times 10^{-11},$$

a fringe displacement completely unobservable even by modern tools.

We want to stress that the estimated degree of breakdown of LLI ensuing from our experiments is in agreement with the existing limits[(2)]. A detailed discussion of this point is given in ref. [12]. Here, we confine ourselves to summarizing the main results.

We recall that two different kinds of LLI violation parameters δ exist: isotropic (essentially obtained by means of experiments based on the propagation of e.m. waves, *e.g.*, of the Michelson-Morley type), and anisotropic parameters (obtained *via* experiments of the Hughes-Drever type[(2)], which test the isotropy of the nuclear levels). The smallest upper limit obtained in the former case is[(2)] $\delta < 10^{-8}$, whereas the upper limits on the anisotropic

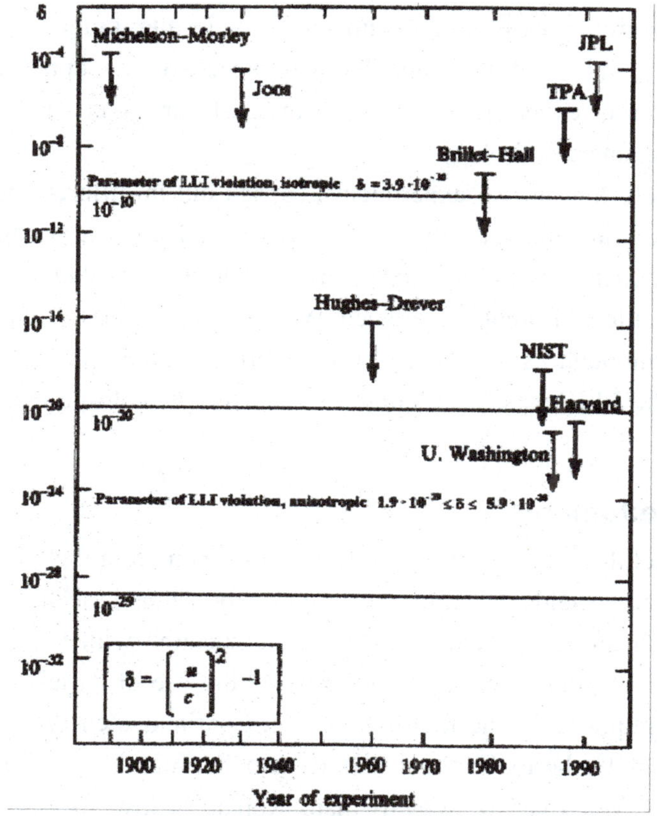


**Fig. 4**. The present experimental situation of the limits on the LLI breakdown parameter $\delta$ (adapted from Will, ref.[2], p.322). The three horizontal straight lines are the limits obtained in the present experiments. See the text.

parameter range from $\delta < 10^{-18}$ of the HD experiment to $\delta < 10^{-27}$ of the Washington experiment(2). In either case, one has to consider, for the evaluation of $\delta$, a *phenomenological LLI invariance breakdown speed v* (*e.g.*, the speed of a hypothetical preferred frame), such that the new speed of light is $u = c + v$. Notice that $u$ is nothing but the *"maximal causal speed"* of the electromagnetic interaction, in Deformed Special Relativity(3), or *"maximum attainable speed,"* in the words of Coleman and Glashow(4,5).

In our framework, the effective LLI breakdown speed $v$ is given by the value found above. Then, it is possible to show that the isotropic LLI parameter corresponding to our effect has the value(12) $\delta \cong 4 \times 10^{-10}$, which is lower by two orders of magnitude than the upper limit for the isotropic case.

In the anisotropic case, the parameter $\delta$ is in the range[12] $2 \times 10^{-29} < \delta < 6 \times 10^{-20}$, and therefore compatible with the anisotropic upper limits.

The present experimental status of the LLI parameters, in light of our results, is summarized in Fig. 4.

In conclusion, in two experiments, carried out in different places, with different experimental set-ups, we observed an effect of a voltage induced across a conductor by a *stationary* magnetic field that could be interpreted as a violation of local Lorentz invariance. We have at present no sound explanation for it, although its parameterization in terms of an effective speed yields values of the LLI breakdown parameters consistent with the existing upper limits.

## Acknowledgments

F.C. is grateful to ing. Salvatore Di Loreto for his help in the measurements of the magnetic fields, to Antonio Centonza for valuable technical support, and to the Institute "Leonardo da Vinci," associated with the Engineering Faculty of L'Aquila University, for allowing the use of its laboratories and equipment required for the first experiment. Continuous encouragement and support from Umberto Bartocci and Renato Scrimaglio are gratefully acknowledged. Moreover, we warmly thank Roman Jackiw, for his kind invitation to deliver a seminar at MIT (held by F.C. on May 8th) and for his appreciation of our work, and Sidney Coleman, Bruno Coppi and Sheldon Glashow for their kind interest and useful discussions.

## References

1. See *e.g.*, Bjorken, J.D. & Drell, S.D., *Relativistic Quantum Fields* (McGraw-Hill, N.Y., 1965), sect.11.1.
2. Will, C.M., *Theory and Experiment in Gravitational Physics* (Cambridge Univ. Press, rev.ed.1993), and references therein.
3. Cardone, F. & Mignani, R., *Found. Phys.* **29**, 1735 (1999), and references therein.
4. Coleman, S. & Glashow, S.L., *Phys. Lett. B* **405**, 249 (1997).
5. Coleman, S. & Glashow, S.L., *Phys. Rev. D* **59**, 116008 (1999), and references therein.
6. Jackiw, R., *"Chern-Simons violation of Lorentz and PCT symmetries in electrodynamics"* (hep-ph/9811322, 13 Nov. 1998), and references therein.

7. Cardone, F. & Mignani, R., *"On a nonlocal relativistic kinematics"* (INFN preprint n.910, Roma, Nov. 1992).

8. Cardone, F., Francaviglia, M. & Mignani, R., *Gen. Rel. Grav.* **31**, 1049 (1999), and references therein.

9. Cardone, F. & Mignani, R., *Int. J. Modern Phys. A* **14**, 3799 (1999).

10. Alley, C.O., *"Proper time experiments in gravitational fields with atomic clocks, aircraft, and laser light pulses,"* in *Quantum optics, experimental gravity, and measurement theory* (P.Meystre and M.O.Scully eds., Plenum Press, 1983), p.363.

11. Bartocci, U., Cardone, F. & Mignani, R., *"Looking for a possible breakdown of local Lorentz invariance for electromagnetic phenomena: Theory and first experimental results"* (*Found. Phys. Lett.*, in press).

12. Cardone, F. & Mignani, R., *"Observation of a DC voltage induced by a steady magnetic field: A possible electromagnetic breakdown of Lorentz invariance?"* (submitted for publication).

13. Cardone, F. & Mignani, R., *"Experimental evidence for a voltage induced by a stationary magnetic field: A possible breakdown of Lorentz invariance"* (submitted for publication).

14. Heald, M. A., *Amer. J. Phys.* **52**, 522 (1984).

15. Bartocci, U. & Mamone Capria, M., *Am.J.Phys.* **59**, 1030 (1991); *Found. Phys.* **21**, 787 (1991).

16. Assis, A.K.T., Rodrigues Jr., W.A. & Mania, A. J., *Found. Phys.* **29**, 729 (1999).

17. Jefimenko, O. D., *Amer. J. Phys.* **30**, 19 (1962).

18. Sansbury, R., *Rev. Sci. Instrum.* **56**, 415 (1985).

19. Edwards, W.F., Kenyon, C. S. & Lemon D. K., *Phys. Rev. D* **14**, 922 (1976).

20. Maglic, S. & Bartlett, D. F., *Rev. Sci. Instrum.* **61**, 2637 (1990).

21. Lemon, D. K., Edwards, W.F., & Kenyon C. S., *Phys. Lett. A* **162**, 105 (1992).

22. See Combourieu, M.C. & .Vigier, J.P, *Phys. Lett. A* **175** (1993), and references therein.

23. For a theoretical discussion, see *e.g.,* Thorne, K.S., in*: 300 Years of Gravitation,* eds. S.W. Hawking and W. Israel (Cambridge Univ. Press, Cambridge, 1987). Recent experimental upper limits are given in P. Astone *et al.*, *Phys. Lett. B* **385**, 421 (1996).

Made in the USA
Monee, IL
04 November 2026

67377660R00105